PRESTON CHAR

A MATTER

OF SCALE

UNTANGLING THE TITANIC CHALLENGE OF
HUMANITY'S CLEAN ENERGY FUTURE

RIVER GROVE
BOOKS

Published by River Grove Books
Austin, TX
www.rivergrovebooks.com

Distributed by River Grove Books

Design by Greenleaf Book Group and Brian Phillips
Cover design by Greenleaf Book Group and Brian Phillips

Cover images copyright Fourleaflover, tr3gin, & Tang Shizhen. Used under license from Shutterstock.com

Information in Figures 19 and 31 is courtesy of the North American Electric Reliability Corporation's website is the property of the North American Electric Reliability Corporation and is available at https://www.nerc.com/pa/RAPA/ra/Reliability%20Assessments%20DL/ Special%20Report%20-%20Accommodating%20High%20Levels%20 of%20Variable%20Generation.pdf. This content may not be reproduced in whole or any part without the prior express written permission of the North American Electric Reliability Corporation.

Publisher's Cataloging-in-Publication data is available.

Print ISBN: 978-1-63299-383-0

eBook ISBN: 978-1-63299-384-7

First Edition

This book is dedicated to my loving parents,
Martin Charles and Peggy Jean Urka.

CONTENTS

PROLOGUE: Ryan Boyle's Question vii

PREFACE: Climate Change and Low-Carbon Power ix

CHAPTER 1: High-Carbon Power 1

CHAPTER 2: Policy Questions 7

CHAPTER 3: Energy School 19

CHAPTER 4: Electricity School 29

CHAPTER 5: Titanic Scale 61

CHAPTER 6: Hydropower 71

CHAPTER 7: Geothermal Power 77

CHAPTER 8: Nuclear Power 79

CHAPTER 9: Wind Energy 91

CHAPTER 10: Solar PV Energy 97

CHAPTER 11: Concentrated Solar Energy 101

CHAPTER 12: Below Utility-Scale Low-Carbon Choices 105

CHAPTER 13: Societal Valuation of Utility Power Sources 111

CHAPTER 14: Policy Actions 137

CHAPTER 15: Applying the Model 143

CHAPTER 16: Conclusion 151

EPILOGUE: My Answer to Ryan Boyle 153

APPENDIX A: Glossary 155

APPENDIX B: Powering the Flatlands of Belgium 159

APPENDIX C: Synthetic Fuels 161

APPENDIX D: Storage 165

APPENDIX E: The Value of Predictability 181

APPENDIX F: Nuclear Power Concepts 191

APPENDIX G: Net-Zero Metering 213

APPENDIX H: The Electric Grid Is Not Like the Internet 217

ACKNOWLEDGMENTS 219

NOTES 221

ABOUT THE AUTHOR 241

~~~~~~~~~~ PROL

RYAN BOYLE'

WHEN WRITING THIS BOOK, I strong-arr
to comment, and to suggest improven
and we met a few days later. He gav
looked me in the eye, and stated, "Yo
make a difference. I have been thinkir
roof. Do you think this is a good idea

Based on this and other ego-crush
pletely failed to communicate my mes
and rewrote most of this book. In it
Ryan Boyle. More importantly, you w
entire global society.

CLIMATE CHANGE AND LOW-CARBON POWER

CLIMATE CHANGE AFFECTS you, me, and our entire society. It will affect your children and grandchildren. Abundant and cheap power is the main technical determinant of our economy's capability and wealth. The manner in which we generate power for our society—predominately burning high-carbon fuels—releases carbon dioxide and other greenhouse gases, worsening climate change.

Our global society must set a goal of using low-carbon energy resources to generate power. Our economies can remain wealthy, vibrant, and capable of expansion if we adopt appropriate policies.

This book explores those low-carbon possibilities and will help you understand the solutions society must adopt—the solutions you must advocate for.

A MATTER OF SCALE

We all know that people too easily move from a familiar understanding to applying the same knowledge in an unknown domain; this is a very human leap. However, this leap does not scale. Possessing a home with a kitchen doesn't provide the capability to take on solving world hunger. Likewise, global electricity use is not something you, I, or any individual can intuitively comprehend.

A young child observing the Sun may conclude that the Sun moves around the Earth. It is only as we grew up and went to school that you and I learned the opposite is true. As it is, it took millennia for humanity to arrive at our current understanding of the solar system. Eventually, society cast away the geocentric model to move to a heliocentric model of the solar system. You must cast away familiar understandings of comforting but misleading conventional wisdom about energy and electricity. To understand energy policy, you must understand the scale of energy that society uses.

Humanity generated 23,696,000,000 megawatt-hours (MWh) of electric energy in 2017,[1] which is 2,703,000 megawatts (MW) of power for each of the 8,766[2] hours in a year. This is a huge, incomprehensible, titanic amount of energy—and electricity is only about 20 to 25 percent of the world's total energy[3] use!

Meet a well-muscled weightlifter at the gym, and you think to yourself, *This fellow is quite strong.* You know well within your own experience that the ability to bench press 130 kilograms (286 pounds) is quite amazing. You then meet Lasha Talakhadze, an Olympic-class weightlifter who can snatch 220 kg (484 pounds) and clean and jerk 258 kg (568 pounds; i.e., lifting three fair-size men over his head). His arms are the size of other people's legs! His legs are the size of other people's waists! He is really, really strong. When you look at Lasha Talakhadze, you have a reference point to understand strength.

Then you meet Atlas, strongest of the Titans, holding up the Heavens themselves. Your reference points of strength, yourself, others, and even Lasha Talakhadze, are completely meaningless. We have no method of measuring the full strength of Atlas, but we can be sure it is titanic! This is the magnitude that we must come to grips with.

CONTENTS

PROLOGUE: Ryan Boyle's Question vii

PREFACE: Climate Change and Low-Carbon Power ix

CHAPTER 1: High-Carbon Power 1

CHAPTER 2: Policy Questions 7

CHAPTER 3: Energy School 19

CHAPTER 4: Electricity School 29

CHAPTER 5: Titanic Scale 61

CHAPTER 6: Hydropower 71

CHAPTER 7: Geothermal Power 77

CHAPTER 8: Nuclear Power 79

CHAPTER 9: Wind Energy 91

CHAPTER 10: Solar PV Energy 97

CHAPTER 11: Concentrated Solar Energy 101

CHAPTER 12: Below Utility-Scale Low-Carbon Choices 105

CHAPTER 13: Societal Valuation of Utility Power Sources 111

CHAPTER 14: Policy Actions 137

CHAPTER 15: Applying the Model   143

CHAPTER 16: Conclusion   151

EPILOGUE: My Answer to Ryan Boyle   153

APPENDIX A: Glossary   155

APPENDIX B: Powering the Flatlands of Belgium   159

APPENDIX C: Synthetic Fuels   161

APPENDIX D: Storage   165

APPENDIX E: The Value of Predictability   181

APPENDIX F: Nuclear Power Concepts   191

APPENDIX G: Net-Zero Metering   213

APPENDIX H: The Electric Grid Is Not Like the Internet   217

ACKNOWLEDGMENTS   219

NOTES   221

ABOUT THE AUTHOR   241

RYAN BOYLE'S QUESTION

WHEN WRITING THIS BOOK, I strong-armed many of my friends to proofread, to comment, and to suggest improvements. Ryan Boyle duly read the book, and we met a few days later. He gave me his feedback and then paused, looked me in the eye, and stated, "You don't tell the reader how they can make a difference. I have been thinking about installing solar panels on my roof. Do you think this is a good idea? Will this help stop climate change?"

Based on this and other ego-crushing feedback—given that I had completely failed to communicate my message—I left our meeting, went home, and rewrote most of this book. In it, you will find my specific answer to Ryan Boyle. More importantly, you will also find my general answer to our entire global society.

FIGURE 1. Lasha Talakhadze in Rio 2016. Atlas makes this amazingly strong fellow look puny.

FIGURE 2. Atlas, the strongest Titan.

When thinking about 23,696,000,000 MWh of electricity energy, we must think in terms of titanic scale, not the mere puny, human, ordinary, everyday measures. This is a lot of energy—so much that our everyday perspective is quite meaningless. For example, we can make the following conversions, and few people will understand better.

- 1,310 Grand Coulee Dams, the largest hydroelectric plant in the United States

- 2,159,680,000 US homes' consumption;[4] if we could build them, that would be 7.4 homes per American

- 270 Belgiums'[5] worth of consumption

Who understands the generation power of one Grand Coulee Dam? Can you understand how much energy your home—or even a single light bulb—consumes? If you understand how much electricity your home uses, can you imagine over two and a half billion of the same? Imagining the multiples above may be possible for a Belgium electric grid manager, but the rest of us have no hope of comprehension. Comparisons made with the electricity demand of Belgium are useless, except to point out that if a solution is inappropriate for Belgium, then is it appropriate anywhere else?

Looking at these ridiculous and meaningless comparisons, you quickly realize it is enough to understand this is an awesome amount of electricity. This Atlas-size problem requires titanic thinking about the solution!

AUTHOR-READER CONTRACT

I will try to not insult the reader by breaking units down into number of light bulbs per person or Empire State Buildings of volumes or other meaningless comparisons unless, as above, the point is to show the unit is incomprehensibly titanic.

In return, I ask you to keep the scale of these solutions in mind and not indulge in unbelievable fantasies about the world solving these problems through conservation by unplugging phone chargers or distributed generation by putting a solar panel on your roof. One thousand MW hydropower is titanic scale; unplugging phone chargers and installing household solar do not compare to the strength of Lasha Talakhadze, let alone Atlas.

As it is, only 12 percent of the average rich world person's energy consumption is due to residential electricity, and 18 percent of energy consumption is due to residential heating.[6] You may be able to replace the energy used for heating, cooling, cooking, hot water, and light, but you will not be able to replace the electricity that pumps water to your house and treats the sewage exiting your house or the fuel used in picking up your garbage. Your food was planted, irrigated, harvested, shipped, processed, and packaged using energy you do not control. Your employer may require you to burn fuel to travel daily to an office in which you have no control over the thermostat or lighting. The buildings you use are constructed with cement, and you will starve without industrial ammonia.[7] You must move past the common understanding of the energy you consume daily and attempt to grasp the entirety of the energy needed to support our society.

Unplugging your phone charger won't even reduce your electricity bill. Putting solar panels on your home may reduce your electricity bill, but it won't solve our societal problem. Putting solar panels on each and every home will at best address 12 percent of our problem, but it also won't solve our societal problem.

This book is for laypeople, and as such, the science and engineering are simplified. Such nuances are left for the reader's further study; therefore, the endnotes often contain a reputable website for the curious. The calculations are rounded, and significant digits are ignored. This book is written favoring clarity over quibbles. Although data from different sources, time periods, or geographies differ, and the calculations are made on the back of an envelope, the basic conceptual meaning presented holds true. If more recent data are used, calculations carried out to higher precision, and accurate rates of growth used, the end result is that a few decimal points will have shifted insignificantly. The insights will remain the same.

Energy production and consumption are at the vast scale of our entire society. Each of us individually can take insignificant steps toward our low-carbon power goal, but together we can address this goal and implement a low-carbon power solution.

STANDARD UNITS

Throughout the book, standard units will be used:

> Power: megawatts (MW)
> Energy: megawatt-hours (MWh)
> Power density: watts per square meter (W/m^2)
> Greenhouse gas intensity: grams carbon dioxide equivalent per kilowatt-hour (gCO$_2$/kWh) or tons carbon dioxide equivalent per capita (tCO$_2$/ capita)
> Length: meters (m) or kilometers (km)
> Area: square meters (m^2), hectares (ha), or square kilometers (km^2)
> Mass: kilograms (kg)

CHAPTER 1

HIGH-CARBON POWER

MANKIND'S POWER IS currently created largely using fossil fuels so dirty as to be unsustainable. These dirty fuels release greenhouse gases. These emissions warm the planet, acidify the oceans, and create human health problems. Any of these three problems is a challenge for humanity to solve; all three taken together raise the value of the solution.

I'll refer to these problems as *climate change* and will simply assert that climate change is occurring, is undesirable, may be mitigated by reducing the greenhouse gases produced by humans, and requires a quick solution implemented with extreme urgency. No further effort will be expended to defend these assertions.

FIGURE 3. Ocean acidification. Pterapod shell dissolved in seawater adjusted to the ocean chemistry projected for the year 2100.

GREENHOUSE GASES

Collectively, the term *greenhouse gases* refers to carbon dioxide (CO_2), methane (CH_4), and other gases such as nitrogen oxides and fluorinated gases. These gases create a layer in the Earth's atmosphere that traps radiation from the Sun, warming the Earth just like a greenhouse.

In Figure 4, CO_2-FOLU is the amount of carbon dioxide from forestry and other land use, such as agriculture. Methane is quite a potent greenhouse gas and is a substantial contributor. *Other* is the category of nitrogen oxides and fluorinated gases that make up the remainder of contributors. It is clear that the bulk of greenhouse gas emissions are CO_2 fuels from burning coal, natural gas, and other fossil fuels. Sadly, it is also clear that those emissions are rising.

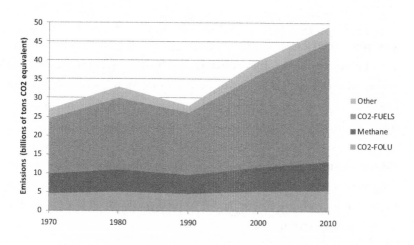

FIGURE 4. Cumulative greenhouse gas emissions.

The effect of greenhouse gas is measured in units of effective CO_2. For example, a single methane molecule has an effect of 84 CO_2 molecules.[1] As greenhouse gas may be measured in units of effective CO_2, and CO_2 requires carbon in its production, the phrase *low-carbon* is used within this book as the antonym of the phrase *greenhouse gases*.

Greenhouse gas intensity, measured in gCO_2/kWh, includes the mining, refining, and transportation of the fuels needed to generate electricity. For example, coal generation's greenhouse gas intensity is mostly due to the fuel itself, but it includes everything necessary to get the coal to the powerplant.

Greenhouse gas intensity also includes the material mining, manufacture, installation, and decommissioning of plant and equipment over projected lifetimes. This applies to coal and natural gas plants, hydroelectric dams, geothermal wells, nuclear reactors, wind turbines, photovoltaic (PV) solar panels, and concentrated solar mirrors.

COAL

High-carbon coal and other fossil fuels[2] are particularly bad because they not only affect climate change but also create other pollutants. Their basic pollutants are these:

- Greenhouse gases: warming the planet and acidifying the oceans
- Soot: microscopic particulates directly impacting human health[3]
- Acid rain: sulfur oxides and nitrogen oxides harming forests and other flora
- Chemical poisons: mercury, cadmium, and other heavy metals within the fuel directly impacting human health
- Radioactive[4] poisons: uranium and other actinides[5] within the fuel directly affecting human health

FIGURE 5. German coal power plants kill 4,350 people per year. Global coal power plants cause death and disease for millions.[6]

Coal power plants, when operated properly and according to the law, kill people by causing asthma, cancer, and pulmonary disease.[7] Without reservation, these plants must be replaced with low-carbon alternatives.

NATURAL GAS

High-carbon natural gas is categorized separately from higher-carbon coal, as it is a cleaner-burning fuel with approximately half the emissions of coal. Natural gas is nearly but not quite 100 percent methane; the proportion is relatively small and therefore not considered further here. A further simplification is that the ease of controlling methane combustion results in controllable emissions of nitrogen oxides. However, piping and burning natural gas still creates quite a lot of greenhouse gas emissions—warming the planet and acidifying the oceans.

However, these plants must be replaced with low-carbon alternatives as fast as possible for three reasons:

1. Natural gas power plants, when operated properly and according to the law, kill people frequently in explosions and accidents.

2. Half as bad as coal is not an endorsement of natural gas! The combustion of natural gas is a major contributor to greenhouse gas emissions.

3. Methane itself, if it leaks unburned, is a more powerful greenhouse gas than carbon dioxide. Natural gas leaks from wells, storage tanks, pipelines, and processing plants contribute 3 percent of total greenhouse gas emissions.[8] Simply by stopping the use of natural gas, we can avoid the leaks and immediately drop our greenhouse gas emissions by 3 percent overnight.

FIGURE 6. New York City, 2014 East Harlem explosion, eight dead.

FIGURE 7. San Bruno, California, 2010 pipeline explosion, eight dead.

FIGURE 8. Kaohsiung, Taiwan, 2014 multiple gas explosions, thirty-two dead

CHAPTER 2

POLICY QUESTIONS

DISTRACTIONS

A few red herrings are often posed as potential questions for humanity about lowering carbon emissions. Here are a few of those questions and the reasons we need not consider those issues. What we can do is change how we generate that power and move from the high-carbon fuels of coal and natural gas to low-carbon alternatives.

CAN WE AFFORD A TRANSITION TO A LOW-CARBON ELECTRICITY WORLD?

As the adage goes, "A problem that may be solved with money is not a problem." Having provided the answers to the difficult technical, environmental, and societal questions in the following, the act of allocating and spending the money needed is truly simple. We can afford the transition.

CAN WE REDUCE ELECTRICITY USAGE?

Reduction (as opposed to conservation) implies quotas and rationing, enforcement of lifestyle changes, and establishing generation ceilings to reduce electricity generated. If you lack the appreciation for the scale, you might believe that if millions of people all just change their lifestyle to use less electricity, then we can have a major impact. Now contemplate the fact that on a planet of 7.6 billion, even a million people are a small percentage.[1] Reducing electricity to the last sustainable greenhouse gas emission level would mean reducing power output to pre–Industrial Revolution levels. Such a highly rural and power-poor world cannot support the world's current—let alone future—population.

We simply need to generate more electricity, not less. First, we need to replace existing high-carbon electricity generation with new low-carbon electricity. Second, we need to replace existing high-carbon nonelectric consumption, such as transportation, with new low-carbon electricity. Third, we need to provide low-carbon electricity to those in the world who have no power at all. We cannot ration our way to a low-carbon future.

CAN WE FOCUS ON ENERGY CONSERVATION?

Conservation (as opposed to reduction) implies consuming electricity through better and more efficient appliances, buildings, tools, and industrial processes. Conservation improves the value of all energy resources; even coal becomes more valuable as less greenhouse gas is produced. Policy choices such as taxes on greenhouse gas emissions or programs such as Energy Star[2] to certify the efficiency of appliances are great, but they raise the value of electricity generation; they do not reduce electricity consumption. Also, as we improve the efficiency of residential appliances and industrial processes, each further improvement approaches diminishing returns.

And we need to remember that none of the technologies to generate low-carbon electricity rely on the ability to improve energy conservation. We cannot reach a low-carbon future by simply improving energy conservation.

MUST WE IMPROVE OUR TRANSMISSION AND DISTRIBUTION SYSTEM?

Similar to conservation, improving the electric grid raises the value of all energy resources; even coal becomes more valuable with a better electric grid. An improved grid allows some low-carbon technology choices to achieve economic practicality. It is an open question as to whether improving the grid should be included in the costs of those choices, but we do know those technology choices cannot be made without a better grid. We must improve the grid irrespective of our goal to lower greenhouse gas emissions.

MUST WE SACRIFICE OUR ECONOMY FOR CLEAN ELECTRICITY?

The simplest solution to retaining the same economy using low-carbon electricity is to provide low-carbon electricity at the existing prices. If low-carbon solutions cost the same as coal, nothing is lost. It is easier to see this relationship if the externality costs of greenhouse gas emissions are accounted for explicitly.[3] Society becomes wealthier as more power is consumed and poorer as greenhouse gas pollutes. A carbon tax that accounts for the cost of greenhouse gas pollution can be set to make explicit the value being destroyed through greenhouse gas pollution. With lots of low-carbon power, we can have a strong economy, driven by higher power consumption and lower greenhouse gas pollutants. We strengthen our economy by using lots of low-carbon electricity.

RELEVANT ISSUES

Our society needs to answer the following six policy questions:

1. The technical question: How do humans create electricity using low-carbon resources?

2. The risk question: How does humanity ensure a positive outcome of low greenhouse gas pollution?

3. The environmental question: How do humans create electricity while conserving biosphere diversity?

4. The timing question: Can humanity move to low-carbon electricity within the time period we have?

5. The political question: How do governments create incentives to promote low-carbon electricity?

6. The societal question: How does society create electricity for all mankind?

THE TECHNICAL QUESTION

Several known low-carbon technologies exist that may reduce greenhouse gas pollution by replacing high-carbon fuels. What choices do we have to generate low-carbon electricity? How are we to value those choices?

We are able to reduce emissions by generating low-carbon electricity. Closing high-carbon coal and high-carbon natural gas plants or reducing their output will result in lower greenhouse gas emissions. To replace the electricity, we must then generate more power from low-carbon resources. This prediction has been tested empirically in France, Sweden, and the Canadian province of Ontario. These geographies burn little coal or natural gas and have very low greenhouse gas emissions from their electricity generation. The negative impact of the prediction has been tested empirically in Australia, Germany, China, and the United States. These geographies continue to burn coal and natural gas and have high greenhouse gas emissions from their electricity generation.

THE RISK QUESTION

We must change the way our society generates electricity to a low-carbon method, we must do so quickly, and we must do so with large investments. When betting the future of humanity on an outcome, the solutions must be practical and realistic. Existing, proven, and workable technologies have the advantage.

We must not make irresponsible claims that innovation will come to our rescue over the next thirty years. Now, it may come to be that technology—particularly in energy storage and demand management or perhaps in carbon capture and storage or fusion—does make such progress, and that would be welcome. However, to presume such progress will occur within a given time frame is a risky gamble; it is not a planning parameter for global greenhouse gas emissions reduction.

"I'm willing to bet on 84 years of time to innovate to make storage, things we didn't talk about before, space-based solar entirely realistic, there is a huge opportunity for base-load clean power to get far along."

−DR. DANIEL "HIGH ROLLER" KAMMEN, OPTIMISTIC FAN OF THE BIG GAMBLE[4]

. . .

"I have long understood that losing always comes with the territory when you wander into the gambling business, just as getting crippled for life is an acceptable risk in the linebacker business. They both are extremely violent sports, and pain is part of the bargain. Buy the ticket; take the ride."

−HUNTER S. THOMPSON, GAMBLER[5]

What are the stakes? If Dr. Kammen wins his wager, we can provide him with general acclaim, prizes, money, and the best tables at all restaurants. If he loses, what can he provide the rest of us other than a planet suffering from climate change? There may be many people on this planet willing to buy the ticket and take the ride, even when it is understood that humanity may be on the losing end, ready and willing to join Dr. Kammen's gamble. Does this mean the majority of us should too? We must all be wary of so-called solutions wherein fundamental advances in technologies must occur for success to be achieved. There is no payout greater than the planet's biosphere to gamble for. Therefore, I am not willing to bet the entire world.

THE ENVIRONMENTAL QUESTION

The value of preserving biodiversity on Earth is self-evident. Examples of how a diverse biosphere contributes to human wealth and health are shown by the great advances in medicine and agriculture.

There are three main competitors for land and sea, each of which must compete against each other and for which area is a constraining factor:

1. Agriculture—both food and resource (such as timber) crops
2. Wild—reservations set aside for wild plants to grow and wild creatures to roam
3. Energy—direct energy capture systems such as solar PV or growing biofuels

Unique to the wild is the ability to reduce greenhouse gas by locking away carbon in repositories such as old-growth forests, ancient peats, and undisturbed tundra. Cities and resource extraction represent a very minor amount of area. This diversity value appears self-evident and with little dispute. In order to preserve this value, the constraint of a minimal area devoted to energy farming is recognized. The less area spent on generating electricity, the more that can be devoted to agriculture or the wild.

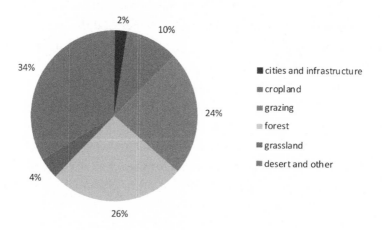

FIGURE 9. Global land area use.[6]

In particular, the more area we devote to the wild, the more greenhouse gas is removed from the atmosphere. For example, trees grow relatively fast, and they remove carbon dioxide for decades or even centuries.

THE TIMING QUESTION

Reducing carbon emissions to sustainable levels by the year 2050 is set as the goal to reduce the impact of climate change to merely tolerable. Although the year 2050 is somewhat arbitrary, the basic timeline of drastically reducing carbon emissions over just the next few decades is critical to our society, our economy, and our environment. It is possible to affect our emissions within this timeline. Extending the period beyond 2050 shifts the world's climate change problem from extremely painful to catastrophic.

THE POLITICAL QUESTION

Governments can encourage low-carbon power with economic incentives. They can support and encourage low-carbon research, build low-carbon test plants, and subsidize low-carbon commercial pilot plants. Governments can streamline planning and permitting of commercial low-carbon plant construction. They can also discourage high-carbon technologies with economic disincentives—raising taxes, imposing fees, and creating diminishing quotas for high-carbon electricity generation. However, the most important task of government with respect to energy policy is to use science to guide society with the important questions and the correct answers.

THE SOCIETAL QUESTION

"When a nation whose welfare is highly dependent on technology
makes vital technological decisions on the basis of political philosophy
rather than on the basis of science, it is in mortal danger."

–DR. BERNARD L. COHEN

Power for all is a good thing, enabling a more efficient economy and, therefore, a wealthier economy. Wealthier economies tend to stabilize to lower birthrates, naturally limiting the population, and enabling an economy with even greater per capita wealth than an economy with a higher birthrate.

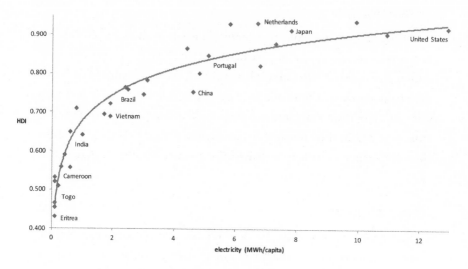

FIGURE 10. High electricity consumption stimulates high human
development index (HDI) outcomes.[8]

Wealthy economies with high electric power consumption have
compact dense agriculture, permitting more land and sea area to be conserved as wild area. Urban and resource extraction areas can be made even
smaller through the greater urban density and resource recycling abundant power permits.

Economic growth and increasing energy use are positive forces that
lead to a stable population wherein more education, capital, and value
are invested in each person. Economic growth provides the capital to
finance low-carbon power, accelerating the reduction of greenhouse
gas. Sophisticated economies are the economies where gross domestic
product (GDP) and power diverge—the higher the GDP, the more efficient use of energy is made—in turn leading to more capital available to
finance low-carbon power.

A policy with the objective of high usage of low-carbon power permits the substitution of electricity for resources. With abundant power,
high levels of recycling replace extraction, power-intensive agriculture
increases carbon-negative wild area, and cheap transport avoids duplication of manufacture.

Power for All Equals Wealth for All

Wealth is heavily correlated with electric power. However, electric power is a causal wealth creator, not merely a correlation. Power-driven pumps bring fresh water and take away sewage. Power-driven factories produce goods and recycle waste. Power provides lighting for the service and education sectors. It is simply not possible to create a wealthy society without a power source beyond the resources of animal muscle or burning wood and dung biomass.

Society needs a high-power resource to become wealthy and to remain wealthy, and it needs to be a low-carbon resource, such as a hydropower, geothermal, nuclear, wind, solar PV, or concentrated solar plant to retain a proper environmentally sound biosphere on the best and only planet in the solar system for human society.

Wealth improves several qualitative and quantitative measures of life among the world's population:

- People are able to lead more fulfilling and happier lives.
  - better water supplies
  - better sewage services
  - better education services
  - better quality food
  - better quantity food
- Birth rates decrease.
  - raising per capita wealth
  - lowering human ecological pressure
  - raising per capita education
- Mortality rates decrease.
  - raising quality of life
- Wealth increases as each of the preceding improvements raise productivity.

The refrigerator is a great example of more power increasing wealth. By storing food safely for long periods of time, people can spend less time being sick, less time shopping, and get more value from their food purchases. People use their extra time earning money, pursuing hobbies, educating themselves, and enjoying life.

In order to achieve poverty reduction, the world electricity supply must increase.

The Bottom Line

The basic three-part answer to the societal question is:

- It is beneficial for people to use more power—improving quality and length of life.
- More power reduces the impact of humans on the planet.
- People will use more power whether we wish them to or not.

You are encouraged to view *The Magic Washing Machine* by Hans Rosling.[9] He sets out, in very human and meaningful terms, what power means to the world.

FIGURE 11. Electric power permits people to get past unproductive, uninspiring work and on to value-added tasks.

FIGURE 12. A lack of electric power drives people to drudgery.

THE LOW-CARBON ELECTRICITY AXIOMS

At this point, you should understand the following:

· High-carbon coal and natural gas must be replaced with low-carbon technologies.

· The low-carbon technologies must exist and be practical to avoid risk.

· The low-carbon technologies must occupy minimal area to reduce environmental impact.

· It is desirable to implement low-carbon technologies by 2050.

· Governments can encourage low-carbon power and discourage high-carbon technologies.

· All humanity must have access to power.

In addition, you should also note that we can afford a transition to a low-carbon power world. We cannot simply reduce total power to meet our greenhouse gas intensity goals. Conservation improves economics and practicality, but it does not reduce greenhouse gas emissions. Finally, we can improve our economies and our societies by using clean power—and lots of it.

CHAPTER 3

ENERGY SCHOOL

TOTAL ENERGY IS THE energy from across all generators (coal, oil, natural gas, hydro, nuclear, geothermal, wind, solar PV, and concentrated solar) and across all sectors (e.g., commercial, transportation, industry, electricity).[1] This can lead to some confusion because electricity is sometimes counted as a separate sector and sometimes embedded with a sector's energy usage.[2]

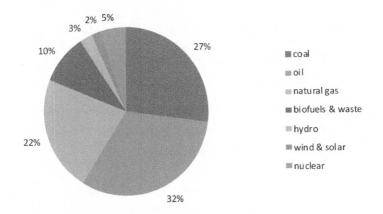

FIGURE 13. Total energy by generation source.[3]

Of course, we care little about how energy is generated, but we do care
a great deal about how large the greenhouse gas emissions are from that
generation. Although natural gas is a cleaner-burning fuel than coal, it is
quite clear that it is a large and significant contributor to greenhouse gas
emissions. The low-carbon energy sources of hydro, nuclear, geothermal,
wind, and solar altogether make up 1 percent of greenhouse gas emissions.

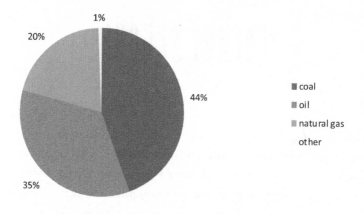

FIGURE 14. Total energy greenhouse gas emissions by source.[4]

We can learn the following basic points from Figure 14:

· Low-carbon technologies are truly low carbon.

· Coal is the worst greenhouse gas emissions offender.

· Natural gas is not a low-carbon technology. Even though it is only
  half as destructive as coal, it remains twenty to forty times more
  destructive than low-carbon technologies.

Breaking out electricity into its own sector, we see that generating electricity is the largest emissions source, followed by transport and industry.
This book concentrates on electricity, with some comments on transport,
industrial process heat, and combined heat and power technologies.

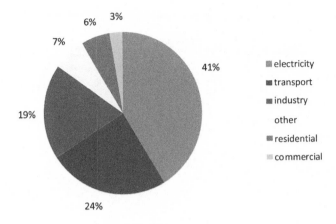

FIGURE 15. Electricity generation is the largest greenhouse gas emitter.[5]

Residential and commercial emissions are relatively low. Thinking at titanic scale, we can quickly see that improvements to 6 to 9 percent of residential and commercial energy through conservation or generation will not have as much impact as improvements to the other sectors.[6] Although such improvements are a worthy goal and will reduce an individual homeowner's or company's bill, the big societal changes that are needed are in utility electricity, transport, and industry.

ELECTRICITY

One-quarter of the world's total energy use is electricity, but this sector is the largest emitter at 13,603,000,000 tons (41 percent) of greenhouse gases; this is unsurprising because over 60 percent of electricity is generated from fossil fuels. The path to reduce greenhouse gas emissions from this sector is straightforward: Stop generating electricity from fossil fuels, starting with coal and ending by removing natural gas.

Biofuels and fuels made from waste may be used in transport, where fuels are more usefully employed. At 2 percent, such a diversion will not significantly affect electricity generation.

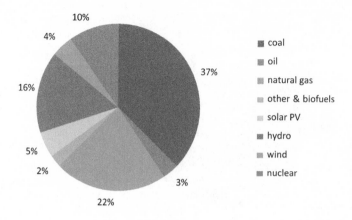

FIGURE 16. Electricity generation by source.[7]

TRANSPORT

Transport is the second largest emissions sector, emitting 7,870,000,000 tons (24 percent) of global emissions. Transport can be electrified to a certain degree. Reducing greenhouse gas emissions by electrifying transport significantly increases the amount of low-carbon electricity that must be generated. Essentially, we need to double electricity generation and more to electrify transport.

Both heat and electricity can also be used to produce synthetic chemical fuels—hydrogen, ammonia, and carbon-neutral methane, alkanes, and diesels from the atmosphere. Low-carbon, high-density chemical atmospheric fuels, rather than fossil fuels mined from the earth, provide for sustainable airplane and heavy vehicle transportation.

HEAT IS USEFUL

Simple, raw heat is a major component of energy use within industry, residential, and the other sectors. Almost all energy used for industry is as process heat—from coal, natural gas, or electricity. Most energy in the home is heat for the building, for cooking, and for water. Most heat energy is produced directly from coal or natural gas.

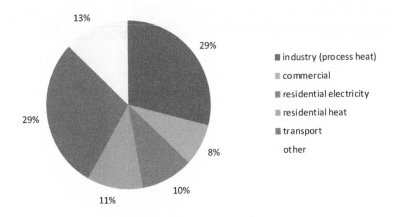

13%

29%

29%

8%

11%

10%

- industry (process heat)
- commercial
- residential electricity
- residential heat
- transport
- other

FIGURE 17. Heat is nearly 40 percent of energy use.[8]

Using electricity to generate heat directly is inefficient, because energy is lost at each conversion step.[9] As we shall learn, to produce heat, the most suitable technologies are geothermal, nuclear, and concentrated solar. Hydro, wind, and solar PV are the least suitable, because their direct output is electricity.

INDUSTRIAL PROCESS HEAT

Industry in total accounts for about 29 percent of energy use and 19 percent of global greenhouse gas emissions. Almost all of that energy and those emissions are to generate process heat; only a minor amount is to power machines.

Process heat is the direct use of the heat, such as raising the temperature of catalysts and chemical processes by placing the reaction vessels and piping in a bath of hot coolant. Industrial processes, such as cement manufacture, ammonia production, and hydrogen generation, each have a high carbon dioxide by-product. Decarbonization of these sectors needs continuous high-temperature process heat. In a double win, direct process heat skips past energy losses from an induction furnace converting electricity into heat and greenhouse gas emission losses from the combustion of natural gas for heat.

Process heat application	Temperature range
Cement production	800–1,450°C
Ammonia production	400–500°C
Electrolysis to produce hydrogen	850°C
Aluminum smelting	940–1,200°C
Glass production	500–600°C
Silica glass production	1,000–1,500°C

TABLE 1. Common process heat applications.

Industry needs process heat between 250 and 1,500°C, but the more common processes need the higher temperatures above 500°C, as is listed in Table 1. Most facilities reach these temperatures by burning natural gas.

Of course, one can produce heat using electricity, but it is almost always cheaper and more efficient to use raw heat. Put another way, to produce heat using electricity will mean doubling electricity generation yet again; using heat directly does not require new electricity generation.

COMBINED HEAT AND POWER

Combined heat and power technologies are also useful because some forms of generation produce low-grade waste heat, on the order of 100°C, which may be used to provide heat to the residential and commercial sectors. Heating buildings is a much bigger slice of global greenhouse gas emissions than electricity for buildings.

Building electricity power plants to produce combined heat and power is efficient and a significant step toward conserving energy use, because we can then take full advantage of that low-grade waste heat.

RESIDENTIAL CONSERVATION AND GENERATION ARE NOT TITANIC SCALE

Spending money, time, and effort on improving the energy conservation of a single building can lower a homeowner's energy demand. Generating electricity through solar PV or heat through concentrated solar can lower a homeowner's energy bill. Installing heat pumps and battery storage can lower the bill still further. However, even across thousands, even millions of homes, this makes little difference to society's greenhouse gas emission problem and low-carbon technology solutions.

We have learned from these graphs that it will take the aggregate of improvements to billions of homes across the world to make significant improvements to the residential sector: Only 6 percent of emissions come from the residential sector, although 21 percent of energy demand is residential. Half of residential demand is heat and half is electricity—11 percent and 10 percent of total demand respectively.

Of course, improving residential emissions is a useful and worthy goal for society to work toward, but it is a goal to be pursued at an incremental pace with local focus. Society and politics should concentrate the main effort on reducing emissions from the utility electricity, transport, and industry sectors. The main push on generation should be on utility-scale generation using low-carbon technologies.

ENERGY IS NOT THE SAME AS POWER

Energy is a measure of the amount of work. Power is a measure of the rate of work, the amount of work over time.[10] From the practical perspective, energy is less useful than power. For example, lightning can transmit a lot of energy in a very short time. Imagine putting a thunderbolt through a light bulb—too much energy in too little time. Although lightning has a great deal of energy, it is not very useful as a power source, because the energy is not spread out over time.[11]

Power is the useful application of energy. It is energy provided at the rate proportional to the needs of the task. Hydroelectricity is a very useful power source for a light bulb, because the generator can provide a calculated, continuous supply of electricity.

	What it is	SI units	Alternative units
Energy	Work or heat	joule (J)	watt-hour (Wh)
Power	Work per unit time	joule per second (J/s)	watt (W)

TABLE 2. Energy may be quoted in watt-hours (Wh) or in joules (J). Power is quoted in watts (W).

To further explore the differences between power and energy, we examine the difference between an ideal 1,000 MW power station and an ideal 1,000 MWh utility-scale battery. The power station provides 1,000 MW continuously; in order to determine the energy produced in that continuous period, multiply the power by the time period.[12] In contrast, the 1,000 MWh battery provides 1,000 MWh in total, and then it is empty.

Generation and storage	Unit size	Peak power	Mode[13] power	Maximum energy provided in first hour	Energy provided in the second hour	Total energy over the year
Continuous power	1,000 MW	1,000 MW	1,000 MW	3,600,000 MJ	3,600,000 MJ	8,766,000 MWh
Fast draining 1-hour energy	1,000 MWh	1,000 MW	0 MW	3,600,000 MJ	0 MJ	1,000 MWh
Slow draining 1-year energy	1,000 MWh	0.114 MW	0.114 MW	411 MJ	411 MJ	1,000 MWh
Drain and fill cycle, 1-hour energy	1,000 MWh	1,000 MW	0 MW	3,600,000 MJ	- 3,600,000 MJ	0 MWh

TABLE 3. 1,000 MW of power generation dwarfs 1,000 MWh of energy storage.

We can see that the ideal, 100 percent efficient power station provides 1,000 MW at the peak, at the mode, in the first hour, in the second hour, and on through the 8,766th hour. At the end of the year, the station providing continuous power will have produced 8,766,000 MWh of energy.

The ideal, 100 percent efficient battery optimized to provide all its energy in one hour provides 1,000 MW of power during this one hour, exactly 1,000 MWh of energy, and then is empty. In the second hour through the 8,766th hour, it provides zero power, because it needs to be recharged.

The ideal battery optimized to provide power over the course of an entire year provides 0.114 MW at the peak, at the mode, in the first through the 8,766th hour. This is continuous power, but it is constrained by the 1,000 MWh of energy the battery was charged with.

The ideal battery, which is charged in one hour and drained in the next hour, does not produce energy. It only changes the time at which the energy is used, resulting in net-zero energy over the course of a year.

When comparing power and energy, you need to understand the time period of the comparison.

- A megawatt of power over an hour has equal energy to a megawatt-hour.

- A megawatt of power over a year has 8,766 times more energy than a megawatt-hour.[14]

- A megawatt of power over a second has 3,600 times less energy than a megawatt-hour.

Comparing energy storage to power generation is tricky. Both the energy capacity and the power rating are needed to evaluate storage. If you know the energy capacity, then you want to know how quickly you can use that energy. You want to know the maximum power rating. If you know the power rating, then you want to know how long it will last. You want to know the total energy capacity.

Of course, a nonideal, real-world battery consumes energy, resulting in net negative energy over the course of a year, because real-world batteries can never be 100 percent efficient when charging or draining. Likewise, a power plant can never be 100 percent efficient, but a real-world power plant can be constructed to produce close to the 8,766,000 MWh desired energy at a continuous 1,000 MW.[15]

Now that we understand something about the various sectors of energy use and the basics of energy and power, we can focus on electricity.

ELECTRICITY SCHOOL

ELECTRICITY IS IMMEDIATELY consumed when generated, and therefore, generation should not produce more electricity than is wanted. Always remember: *It is the consumption of electricity that drives generation. Generation never forces consumption to occur.*

The immediate consumption of electricity is what dictates the desirability of dispatchable power and the undesirability of intermittent energy. As a consumer, when you turn on a lamp, you want a continuous supply of electricity to light it—dispatchable power fits the bill nicely. As a grid operator operating in real time, it is very challenging to accommodate unknown amounts of electricity delivered at unknown times—intermittent energy is a pain in your posterior.

Dispatchable power is what allows the immediate nature of electricity to be practical. As consumption increases, the grid operators dispatch more power. As consumption decreases, the grid operators dispatch less power. It is power—a reasonable amount of electric energy released continuously over time—that keeps your lamp lit.

If your lamp were driven by the *intermittent energy* of thunderbolts captured by a lightning energy farm, then it will explode as an unreasonable amount of electricity enters the lamp. On the off chance that these were very small thunderbolts, your lamp would flicker like mad because the supply of thunderbolts is not continuous.

When you turn on a lamp, the electricity travels immediately from the power plant's generators to the lamp. In order to support the lamp's demand, the generators must ramp up instantaneously to create a tiny amount more electricity. When you turn off the lamp, the generators must ramp down instantaneously to create a tiny amount less electricity.

Of course, the generators are not truly able to react instantaneously, and electricity takes time to travel down wires. To deal with these real-world effects, the generators are always producing a tiny amount more power than needed and are therefore ready to react to your lamp being switched on. The downside is that, while waiting for someone to turn on a lamp and use the extra power, the power plant simply drains away the excess by sending it through useless circuits; the electricity *must* be immediately consumed. Likewise, when you switch off the lamp, that extra, unused energy is simply drained away until the generators compensate to meet the lower demand.

Hydropower produces dispatchable power because it can easily ramp generation up or down, following the demand load. A hydro plant operator chooses when to release the dammed water and when to retain it. A hydro plant can generate the correct amount of energy over time: This is power to accommodate human desire.

In contrast, a wind turbine produces intermittent energy, because the turbine cannot choose when the wind blows. Because wind is intermittent, the wind turbine is not dispatchable. If the wind is not blowing, the turbine cannot generate energy, irrespective of the required demand. If the wind is blowing, the turbine can generate energy, irrespective of any unwanted demand. If the grid is not in balance because the wind is blowing too hard, it simply generates extra energy that must be drained away. It is consumption that drives generation. Shutting down the turbine mechanically to avoid excess energy is called *curtailment*.

Wind turbines create raw energy when the wind blows. However, wind turbines only generate power when the wind blows *and* when someone wants the electricity. This is energy dictated by the weather.

DISPATCHABLE POWER

Dispatch is the request that grid operators put to a generation source in response to changes in the demand load. Dispatchable power may be adjusted to meet the current, instantaneous demand load.[1] Humans decide how much power is needed, for example, by turning on a toaster. Using dispatchable power plants, grid operators need to predict a single variable: the demand load. The following are low-carbon power plant types.

- Hydropower (riverine dams and run-of-river turbines)
- Geothermal power
- Nuclear power
- Storage (although not a generation source, storage is dispatchable)

A hydro plant is dispatchable. The plant's energy is always ready to be converted to electric power on demand. The single unknown is the demand load.

INTERMITTENT ENERGY

Intermittent energy is produced using variable resources to generate electricity. Intermittent energy may not be adjusted to meet the current, instantaneous demand load—by definition it is non-dispatchable—you get what you get. Using intermittent energy farms, grid operators must predict both electricity production and the demand load. Weather conditions determine how much power can be generated, but generation is unresponsive to human command.

The following are types of low-carbon energy farms.

- Utility-scale energy farms, such as:
  - Marine hydroelectric power (wave energy, tidal capture)
  - Wind power (onshore, offshore)
  - Solar photovoltaic (PV) energy
  - Concentrated solar energy

- Microscale generation (nonutility generation, residential, and commercial), such as:
  - Heat pumps
  - Solar PV microgeneration
  - Concentrated solar microgeneration

A wind farm is intermittent. The energy is sometimes ready to be converted to electric power. There must be both wind and demand. The two unknowns are electricity generation and the demand load.

Storage can convert intermittent energy into dispatchable power. Storage mitigates the time of generation to the time of demand but incurs an energy penalty to do so. This simplifies the problem from the two unknowns of electricity generation and electricity demand into the one unknown of the demand load.

Concentrated solar may be considered semi-dispatchable, because its coolant is inherently a thermal storage reservoir, but the energy input—sunlight—is intermittent. Over an hourly time frame, it has sufficient dispatchability, but over a daily time frame, it is intermittent because the next day may have little sunlight and its reservoir would be empty.

DISPATCHABLE POWER VERSUS INTERMITTENT ENERGY

Table 4 summarizes the attributes of dispatchable and intermittent plant types. Neither plant type can predict the demand load, but how does each respond to it? Dispatchability defines the difference between the two plant types. The dispatchable power plant has a known generation amount, which remains stable and consistent. Its output is responsive to the operator based on demand load, maintenance schedules, and price. The intermittent energy plant has a non-dispatchable generation amount. Its output is decided randomly by the weather, the time of day, and the season of the year; the needs of the homes and factories it serves are not taken into account. The plant has neither consistent output nor is responsive to the demand load. This inability to dispatch is due to the intermittency of the energy resource.

	Dispatchable power	Intermittent energy
Dispatchability		
Generation amount	known	unknown
Generation consistency	stable	random
Demand load	responsive	unresponsive
Storage importance		
Over-generation	minimal	random
Under-generation	capacity threshold	random
Value from storage	improvement	necessity
Improved grid		
Production times	scheduled	random
Distance to load	local	unknown
Value from grid	improvement	necessity

TABLE 4. Comparison of dispatchable power and intermittent energy.

All plants may take advantage of storage for over-generated energy, but not all plants require storage. Dispatchable plants rarely over-generate, because they are throttled back when demand is low. If the prediction of peak demand load is accurate, then enough dispatchable plants can be constructed to avoid under-generation or blackouts. The plants simply operate to their maximum-capacity thresholds or lower as determined by the grid. The addition of storage to a dispatchable plant is unnecessary, but it does enable the plant to be run in its most optimal manner, thus resulting in an improvement to the value of the plant.

Intermittent plants randomly over-generate and create waste electricity. Even worse, the over-generation is mostly random. As a result, curtailment—the restriction of generation capacity—is a common daily occurrence for grids with even small proportions of 10 to 20 percent of intermittent energy. Likewise, intermittent plants under-generate in a random manner. When intermittent plants fail to make their contract, the grid operator has no choice but to call on a dispatchable plant to provide reliable electricity. Axiomatically, the grid operator cannot dispatch another separate intermittent plant to respond to the demand load. Intermittent plants operate at the behest of nature, not the instructions of the grid operator. This yo-yo effect between over- and under-generation is disruptive to the grid and a difficult technical problem for the operators and grid systems to deal with. These extreme periods can happen in hours[2] and last weeks, as during the *dunkelflaute*, when there is insufficient sun and wind to generate electricity. The term *dunkelflaute* describes periods of days or weeks that are cloudy and windless during the German winter.

The addition of storage to an intermittent plant is a necessity for the plant to be throttled up and down in a dispatchable manner. Storage is necessary to shave the generation peaks and fill the valleys, creating a more grid-friendly electricity resource.

Similarly, all plants may take advantage of an improved electrical grid, but not all plants require an improved grid. The physical distance to the load matters: The longer the distance, the more electricity is lost.

Dispatchable plants are well-behaved grid citizens: Their starts and stops are scheduled, the physical distances to their loads are local and well understood, and they do not require investments in the grid, although they can take advantage of them. The grid operator can choose to adjust a local dispatchable plant in response to local demand load, because they may bring dispatchable plants online or offline at will.

Intermittent plants are poorly behaved grid citizens: Their starts and stops are essentially random because of the dependence on the season, the weather, and the time of day. The physical distances to their loads constantly change as weather, sunrise, and sunset vary the specific points of generation. The unknown nature of the physical distance to demand load is illustrated by the common occurrence of storms covering hundreds of thousands of square kilometers[3]—raising winds beyond wind turbine design tolerances and cutting off sunlight to solar farms. Because the grid operator is unable to bring intermittent plants online or offline at will, they cannot choose to adjust a local intermittent plant in response to local demand load, but they may be forced to adjust a non-local plant in response to local demand load. Intermittent plants receive the most value from major investments in grid transmission, distribution, and management, because they require these investments in order to be practical.

The major point to remember is that dispatchable power plants have the single unknown of the demand load. Intermittent energy farms have the double unknowns of the random generation times and the demand load.

CURTAILMENT LOWERS RETURN ON ASSETS

Because electricity must be consumed when it is generated, when demand is low, less electricity must be generated. Both dispatchable and intermittent plants must curtail their power generation to match the actual demand. In a dispatchable plant, curtailment is not a big deal; the plant simply lowers generation to demand. In the future, when demand rises, the plant simply produces more.

In an intermittent plant, curtailment is a big deal. In the future, the Sun may be shining, the wind may be blowing, or not. A prediction can be made, but in general, the future is unknowable. However, whether or not the plant has lower revenues due to curtailed generation, the cost of building the plant must be paid. Good weather resulting in over-generation results in curtailment. Therefore, the return on assets of a solar or wind farm is lowered by both bad weather and by weather that is too good.

Solar PV suffers the most from this phenomenon because aggregate generation always reaches a peak at noon across a longitude. Each additional solar PV farm raises the likelihood of curtailment. Therefore, of all generation types, solar PV benefits most from storage installation (changing the time used) or smart grids (changing the amount used) and transmission (changing the latitude used) improvements.

DISPATCHABILITY HAS VALUE

Humans are both diurnal and social creatures. Basically, we do things during the daylight hours, and we do them together. This leads large groups of people, across large geographies, to demand power at the same time. Dispatchable power answers this demand at all times because we can command it to do so. Intermittent energy may or may not coincide with this demand because intermittent resources are determined by the random phenomena of weather.

Each of the following examples illustrate our society's simultaneous desire for electricity. Most of the population uses power at work or school during the day. We use power to cook in the early morning, at midday, and at mid-evening. We use power to heat water for showers or baths in the morning or evening. We use power in the evening for entertainment, such as reading lights or television. Always, we use power to drive the 24/7 background hum of the industrial processes that provide society's material comforts.

Dispatchable power plants have two major economic gains over inter-mittent energy plants. Dispatchability is on demand. The value of a good or service on demand is self-evidently obvious to anybody who has suffered through a lunch with a slow waiter. Given the immediate consumption nature of electricity, this value is all the more apparent. Dispatchability requires fewer plants. Dispatchable plants can produce up to their capacity and take full advantage of new storage capacity and grid improvements. Intermittent plants have random capacity and lower returns due to extra assets: extra plants to provide capacity when generation stops, required storage to provide dispatchable capacity, and needed grid improvements to manage demand load and new electricity flows.

STORAGE ALWAYS INCURS A PENALTY

Storage does not fill itself; it must be filled by an energy source. The amount of energy stored will not be as much as was needed to fill the storage. Storage can never be 100 percent efficient. The type of storage defines its efficiency.

As an example, pumped hydro requires water to be pumped uphill. The energy supplied to the pumps will mostly be stored in the reservoir.

However, the pumps are not perfect and therefore generate heat while operating. The water rushing against pipes experiences friction, and so energy
is lost. When the water is let out through the turbines, more energy is lost
again. Therefore, the efficiency of pumped storage, defined as the energy
recovered from storage over the energy placed into storage, is always less
than 100 percent. Individual installations differ, but typical round-trip
efficiencies for pumped hydro are between 70 percent and 80 percent.

$$\text{Energy}_{\text{recovered from storage}} = \text{Energy}_{\text{placed into storage}} - \text{Energy}_{\text{Friction}}$$

$$\text{Efficiency}_{\text{storage}} = \frac{\text{Energy}_{\text{recovered from storage}}}{\text{Energy}_{\text{placed into storage}}} \times 100\% < 100\%$$

As a second example, recharging chemical batteries require ions to
move across membranes. The battery connections are not perfect and
lose energy through shorts and electrical resistance. The ions bump into
each other and the membrane, the friction producing heat. The heat is all
wasted energy. The temperature change also lowers the efficiency of the
chemical reaction of the ions. During battery discharge, the exact same
inefficiencies lose useful energy a second time. Individual installations
differ, but typical round-trip efficiencies for batteries are 80 to 90 percent
for lithium-ion and 75 to 80 percent for vanadium flow batteries. Table 5
shows how much energy is lost due to cycling batteries every twelve hours
over a year. Of course, we may build physically larger batteries to deliver
the net electricity needed to make up for the efficiency losses, but the bigger batteries still consume the difference between gross and net.

Battery type	Unit size	Energy provided over 1 year	Energy consumed over 1 year	Total energy lost over 1 year
100% efficient ideal battery	1,000 MWh	4,383,000 MWh	4,383,000 MWh	0 MWh
90% efficient lithium-ion	1,000 MWh	4,383,000 MWh	4,870,000 MWh	−487,000 MWh
80% efficient vanadium flow	1,000 MWh	4,383,000 MWh	5,478,750 MWh	−1,095,750 MWh

TABLE 5. Unlike the ideal world, storing energy incurs a penalty in the real world.

Storage can shift the time of electricity consumption, but in exchange, it always reduces the amount of energy delivered.

STORAGE IMPROVEMENTS

We may expect improvements in storage ideas and storage technologies. However, intermittent energy requires that storage be built on a completely different scale. We would need titanic improvements in storage to achieve multiday—let alone multiweek—bulk energy storage capacity.

Pumped hydro, compressed air, and chemical batteries (lithium-ion, vanadium flow, brine, and antimony) have already experienced significant improvements. There are new approaches such as fuel cells, mine-based gravity storage, flywheels, and thermal storage. The fundamental chemistry, physics, and engineering of each of these storage solutions are well known. These solutions may be categorized thus:

- Gravitic: pumped hydro, weights traveling up and down in mineshafts, railcars on mountains

- Mechanical: compressed air, flywheels

- Electrical: superconducting capacitors

- Thermal: thermal storage in coolants, such as molten salt

- Chemical batteries: batteries and fuel cells

- Synthetic chemical fuels: reserved for transport

Each of these classes of storage has a well-developed, centuries-old science underpinning it. Efficiencies in pumps will improve pumped hydro, but will there be a breakthrough in the science of gravity? There is no doubt that lubricants and magnetic bearings will be created to reduce flywheel friction, but will we learn something new about kinetic energies? Superconductors are a relatively new and unexplored space and so have great potential for discovery, but will we create materials that scale charge nonlinearly with their mass and therefore with their volume? Thermal and chemical solutions are restricted by fundamental laws of thermodynamics, but can we expect to learn of substances with tenfold-higher heat capacities or electrochemical potentials with vast differences to create super batteries? Synthetic chemical fuels created from low-carbon electricity will also improve—and likely in big ways. However, the value of these very dense fuels for use in transportation is overwhelmingly more than the value of their use in electricity generation.

The improvements expected will be in efficiency, in density, and in cheaper costs. All of these improvements will be incremental. We also need major improvements in scaling of storage capacity and of construction. The timing of when we will achieve the improvements necessary is a critical and separate issue. We may be able to make some predictions, but the timing of a major breakthrough clearly has a large risk factor. The future is certain to be unknowable!

Storage optimism is addressed at greater length in the appendices.

SMART GRID DEMAND SHIFTING

The smart grid has many attributes, but its major "smartness" value is that it will be able to even out over- and under-generation by shifting in time the demand load consumption rather than storage. Some smart appliances react passively to grid conditions—for example, a refrigerator that examines small changes in the grid's frequency to determine load conditions. Even smarter appliances can be manufactured if the grid can actively issue instructions to them, such as informing the appliance of the current price of electricity.

A smart grid prioritizes demand. Time-critical energy requests are served first, and less critical requests are served later. As an example, we may compare television to refrigerators. The television is time critical[4] and requires the electricity to be supplied during the broadcast being watched. The refrigerator's temperature will not change over a few minutes, due to thermal insulation, and can therefore wait for electricity to be supplied at a slightly later time of lower demand.

In Table 6, we can see that the time before failure drives the power priority.

Appliance	Energy inertia	Power priority	Time before failure (without electricity)	Energy efficiency improvements	Power efficiency improvements
Television	none	critical	milliseconds	minimal	minimal
Refrigerator	thermal reserve	can wait	minutes to hours	minimal	major

TABLE 6. Demand shifting power.

Energy Star (certified by the US Department of Energy) televisions are 20 percent more energy efficient, and Energy Star refrigerators are 30 percent more energy efficient than their brethren, but progress at this point is incremental. Major improvement requires fundamental technology changes, such as was the case in the transition from CRT to flat-screen televisions. Although scientists will develop better insulation and coolants and engineers will design better pumps and heat exchangers, it is rather difficult to conceive of a major *energy* efficiency gain in refrigeration.

However, as Table 6 indicates, the refrigerator is capable of *power* efficiency improvements. We can shift the time of the refrigerator's demand load (by seconds, minutes, or hours) to optimize the power delivered. Therefore, for appliances with energy inertia or that can wait in an energy priority queue, we can improve the power efficiency.

This demand-shifting feature of the smart grid helps dispatchable power plants run at peak efficiency and provides a major economic boost to intermittent energy plants by reducing storage needs.

The downside of building a grid capable of large amounts of demand shifting is that doing so is an extremely challenging effort both technically and politically. Society will always be improving the grid, but the grid will never be smart enough to deal fully with a lack of power from intermittent resources. The smart grid will only be able to mitigate the consequences of intermittency.

CAPACITY AND THE CAPACITY FACTOR

The capacity factor of a power plant is an empirical measure equal to the quotient of the practical capacity and nameplate capacity of the plant design.

$$\text{Capacity Factor} = \frac{\text{Capacity}}{\text{Nameplate Capacity}} \times 100\%$$

The practical capacity of a power plant is measured as the power produced over the course of a year, taking into account maintenance, generation peaks and troughs, and intermittency. For dispatchable nuclear power, with its 93.5 percent capacity factor, this is essentially the uptime divided by time, as the plants are run at near 100 percent output and only stop for maintenance. Intermittent solar PV has only a 24.5 percent capacity factor because no electricity is generated during the night and because daylight will vary between sunny, clear, and bright and cloudy, dim, and dark.

The capacity factor enables the comparison of different electricity generation sources. High capacity factors indicate a highly productive and valuable plant. Low capacity factors indicate a low productivity and therefore less valuable plant. The capacity factor does not indicate the level of intermittency; however, only a low-intermittency resource, dispatchable plant can be designed to have a high capacity factor.[5]

Each individual power plant has its own empirically determined capacity factor. Table 7 shows the average capacity factors for each generation type.[6]

Nuclear	Geothermal	Natural gas	Coal	Hydro	Wind	Solar PV	Concentrated solar
93.5%	74.4%	56.3%	54.6%	39.1%	34.8%	24.5%	21.8%

TABLE 7. Capacity factor by generation type.

More nameplate capacity does not result in a higher capacity factor. A bigger dam does not generate more rainfall. More geothermal wells do not generate more heat per well. A bigger reactor does not fission uranium faster. More or taller wind turbines do not generate more wind. A larger solar panel array or mirror does not generate more sunlight.

ADDRESSING LOW CAPACITY FACTORS VIA REDUNDANCY

Using the example of wind, let us construct an ideal wind farm to deliver on a generation contract of 1,000 MW. In order to convert the practical output of 1,000 MW to the average output capacity of wind technology's 34.8 percent capacity factor (rounded to one-third), we need at least three times the plant. Table 8 makes the idealistic but impractical assumption that the wind at all working sites is not intermittent but steady and even. Likewise, a single site is likely to completely fail when it fails, because the turbines within the same area have approximately the same wind. The intermittency of the first turbine has a high correlation with the intermittency of the last turbine within the same wind farm.

	Choice A single large site	Choice B multiple large sites	Choice C extra site, less risk	Choice D many small sites
Sites	1 site	3 sites	4 sites	10 sites
Failure point	1 site down	1 site down	1 site down	2 sites down
Capacity factor	1/3	1/3	1/3	1/3
Per-site nameplate capacity	3,000 MW	1,000 MW	1,000 MW	400 MW
Total nameplate capacity	3,000 MW	3,000 MW	4,000 MW	4,000 MW
Total average generation	1,000 MW	1,000 MW	1,333 MW	1,333 MW
Total average curtailment	0 MW	0 MW	333 MW	333 MW
Generation post-failure	0 MW	667 MW	1,000 MW	1,066 MW
Impact	blackout	brownout	contract met by 3 sites	contract met by 8 sites

TABLE 8. Installation scenarios addressing low capacity factors.

Separating the plant over a large geographic area with multiple sites minimizes the possibility of failure resulting in a blackout or a brownout. A wind turbine at the first site has a low correlation with a wind turbine at the last site. Moving from choice A to the multisite choices B, C, or D raises reliability. Moving from choice A, a complete blackout on failure, to break even or a brownout with choices B, C, and D also raises reliability.

Choice B works when the multiple sites are anticorrelated; that is, a lack of wind in a site predicts extra wind in another. However, there is no perfect anticorrelation between sites, and so one can have brownouts as generation fails to meet demand or curtailment as energy is not generated. Again, we see the importance of storage in moving energy in time from over-generation to under-generation.

Choices C and D demonstrate that increasing the number of uncorrelated plants raises the reliability by requiring more plants to fail before we cannot meet our generation contract. Choices C and D indicate that many smaller plants may be preferable to a few larger ones. The trick is to actively choose sites that are uncorrelated. When the wind fails in San Diego, it is unlikely to fail in Miami simultaneously.

Choices C and D also demonstrate that multiple sites will require more curtailment when all sites are operating. Building these redundant sites is an overinvestment in generation capacity. This overinvestment is necessary when using intermittent energy farms. Sports stadiums are built for game day, not the average day, despite the lack of a crowd on the average day. For the grid, every day is game day. Construction of an energy farm must be to the maximum demand load, not the average demand nor the predicted demand. Infrastructure must accommodate the biggest demand, whether that is stadium attendance or electricity consumption.[7]

INTERMITTENCY

Intermittency is the natural variability and limited predictability of renewable energy generation and is easily observable in wind, solar PV, and concentrated solar energy sources. Because the energy resource itself is variable, the generation plant is intermittent.

Energy resource	Generation plant	Electricity generated as	Controlled by
stable	dispatchable	reliable power	human decision
variable	intermittent	unreliable energy	random weather

TABLE 9. Dispatchability versus intermittency.

WIND INTERMITTENCY

Wind is a natural phenomenon with random generation capacity. Yes, we can make predictions about the weather tomorrow, but we cannot force the wind to conform to tomorrow's demand load.

As we can see from the following graphs, wind generation is not consistent day to day, nor is it similar to the average wind; it can vary significantly by the hour. If you trace through the empirically determined intermittency of Texas wind, as is shown in Figure 18, you will find that each day's wind energy is very different from the day before and from the day after. The average hourly power, as shown by the black dots, is quite steady over a month.[8]

Caution! The actual power on any given hour of any given day does not at all match the average!

On a given day of the month, such as day 21 at the top of the graph, nearly double than the average is produced, and day 14 at the bottom shows only half the average produced. Days 5 and 19 not only show production both above and below the average, but they also show nearly opposite wind patterns across the day. Each day looks completely different because each day has unique weather.

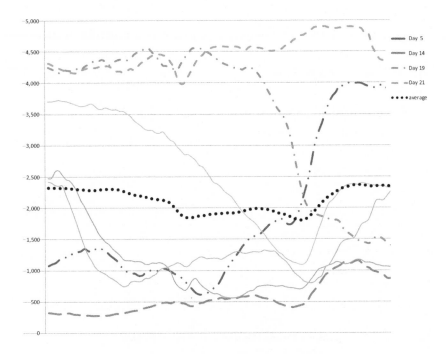

FIGURE 18. Intermittency: Daily wind does not equal average wind.[9]

Figure 18 illustrates a very important point to remember: The daily power values form the average. The average does not produce a constant power value prorated over time.[10] The grid operator cannot rely on this wind farm to consistently produce its nameplate capacity, nor even rely on the farm to consistently produce 34.8 percent of its nameplate capacity. The grid operator learns in real time how much electricity this wind farm produces.

Intermittency is a natural phenomenon that cannot be reduced; it can only be mitigated after the fact using storage or demand shifting via the grid.[11] To reduce wind intermittency, we would need to control the flukiness and strength of wind gusts. Improved weather prediction will enable better *post facto* handling of electricity generation and the demand load, but it will not change the random nature of intermittent energy resources.

As we see from the graphs of Canadian wind in Figure 19, sometimes the weather happens to correlate well with demand, and sometimes it does not.

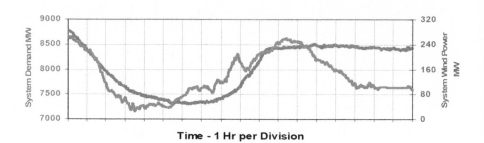

**Alberta System Demand and Wind Power
Do Not Correlate Well Jan 6, 2006**

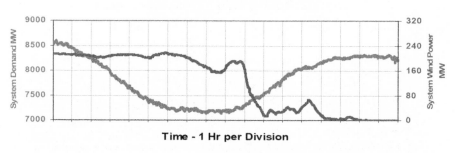

FIGURE 19. Two distinct days showing similar demand but completely different weather.
Variable energy resources are driven by weather, not by human desire.[12]

In addition to hourly fluctuations, wind is strongest in the spring and weakest in the summer.[13] Building out wind turbines to supply the grid requires building for the capacity to match the time of year with minimum generation and maximum demand. Therefore, in spring, generation is over capacity, resulting in a low return on assets from the investment.

SOLAR INTERMITTENCY

As we can see from the graphs of Texas sun in Figure 20, solar generation on any given day is not the same as the average day or as five days before or after. We also see seasonal intermittency across a six-month time frame, when comparing Figure 20 to Figure 21 (the two graphs are at the same scale).

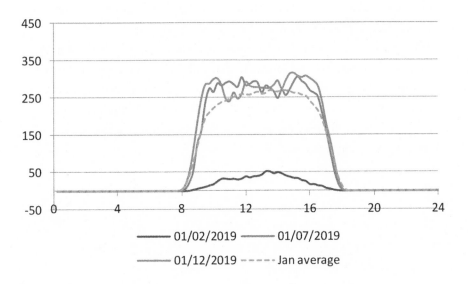

FIGURE 20. Electric Reliability Council of Texas hourly solar (MWh versus hour of day).

During the day, clouds will pass between the panel and the Sun at random times and for random durations. To reduce solar PV or concentrated solar intermittency, we would need to control the trajectories of clouds.

Solar starts generation abruptly in the morning and stops just as abruptly in the evening. Solar PV's nighttime electricity generation will be zero. It is not intermittent during the night; it is zero.

The three days plotted, each five days apart, show hourly fluctuations. This hourly fluctuating intermittency is of a type we may be able to deal with using short-term storage. We also see that days in January appear very cloudy or very sunny. The electricity produced on January 2nd is nowhere near the average for the month; this daily fluctuating intermittency is more difficult to mitigate, given the difficulties of long-term storage.

The start and stop times are clearly season-dependent—because of both the variation in day length and the variation in seasonal weather. The graphs clearly show a longer and much sunnier day in July and a shorter and darker day in January. Solar radiation is strongest in the summer and weakest in the winter. This seasonally fluctuating intermittency is extremely difficult to mitigate because seasonal storage is at the limits of humanity's science and engineering capabilities, perhaps beyond it. A simple solution is to build out redundant solar PV or concentrated solar to supply the grid. Using the redundant solar farms approach requires building for the capacity to match winter generation. Therefore, generation will be over capacity during the summer, resulting in a low return on assets from redundant investments.

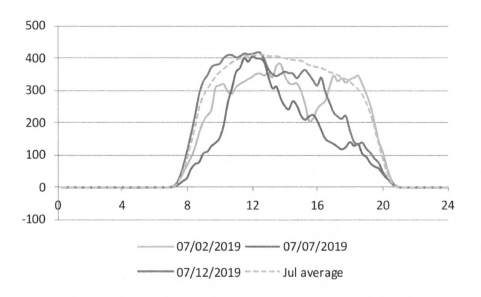

FIGURE 21. Electric Reliability Council of Texas July 2019 Texas solar (MWh versus hour of day).

Solar radiation is a natural phenomenon with random generation capacity. Yes, we can make predictions about the clarity of skies tomorrow, but we cannot force the weather to conform to tomorrow's demand load.

ADDRESSING INTERMITTENCY

There are a few strategies to improve generation reliability of high-intermittency energy resources. Improving generation reliability is beneficial because then the difficult technical and social challenges of demand management will be reduced.

Intermittent energy farms have four mitigation strategies:

1. Storage: building energy storage in order to provide power at a more appropriate time

2. Smart grids: adjusting demand to match generation

3. Natural gas pairing: using high-carbon resources to provide dispatchability

4. Redundancy: overbuilding to ensure that minimum generation can meet maximum demand

There is no free lunch. Each of these strategies requires using energy sources with low capacity factors and high intermittency. We must over-invest in assets that will be idle some part of the time to deliver reliable power. Intermittency's overinvestment results in a low return on assets to ensure reliability.[14]

THE IMPORTANCE OF STORAGE TO INTERMITTENT ENERGY GENERATION

Increased energy storage capacity increases the value provided for both dispatchable and intermittent generation.

Dispatchable plants do not require storage, but their value is improved by sending the relatively small amount of unused energy produced into the useful storage circuit while waiting for lamps to turn on rather than into a useless drain circuit. Dispatchable plants can run closer to their peak efficiency generation modes, which is typically maximum capacity.

Storage is vital to making electricity generation economic for intermittent energy farms. Without storage, when the Sun is shining and demand is low, a relatively large amount of unused solar PV energy will be lost through a useless drain circuit or curtailment. When the day is cloudy and demand is high, solar PV is unable to provide electricity; without storage, a blackout occurs, and with storage, previously saved solar energy keeps your lamp lit. Wind has the same story as solar PV. Concentrated solar has enough storage built in to behave in a dispatchable manner over several hours' duration. However, on the daily timescale, concentrated solar doesn't have enough inherent storage to respond if the day is not sunny.

Most wind and solar PV farms do not build storage appropriate for their size. Instead, these generators export their intermittency problem onto the electricity grid, making it a problem for the grid and other generation plants. Grid operators keep large capacity reserves of coal and natural gas online in order to deal with this intermittency—producing greenhouse gas without any net benefit.[15] Concentrated solar farms inherently build enough daily storage appropriate to their size.

THE IMPORTANCE OF A SMART GRID TO INTERMITTENT ENERGY GENERATION

Any improvement to the grid will improve the value of both dispatchable and intermittent generation. The smart grid can be described as directing the electricity to travel with the least friction from where it is produced to where it is consumed.

A smart grid improves the transmission and distribution of electricity by using real-time, automated, interactive computers to optimize the physical operation of the grid. Smart appliances can optimize their operation with regard to grid conditions by turning themselves on or off. Smart meters allow customers to monitor and understand their electricity usage. Smarter communications enable the grid to make better predictions of demand load, to detect faults and heal them, and to change the grid topology. A smart grid may have pricing linked to the time of day or the current demand load.

Demand management through smart appliances has limitations. A residential dishwasher may delay its washing cycle by a few hours. However, a hungry consumer using a toaster is unlikely to purchase a toaster that fails to cook on demand. Likewise, a busy restaurant quickly cycles through its dirty dishes both to serve its customers and to maintain hygiene. Restaurants are unlikely to purchase dishwashers that fail to wash on demand.

The scale of each grid is itself enormous.[16] Each grid extends across extremely large geographic areas containing hundreds of thousands of kilometers of transmission wire, millions of customers (from single households to large industrial plants), and billions of appliances (from residential refrigerators to industrial induction forges). Making the grid smarter will require upgrades to management given the less predictable nature of intermittent generation, upgrades to demand management across those billions of appliances, those millions of customers, and those hundred thousand square kilometer geographies to better match generation and demand. It will require transmission and distribution extensions as nonlocal intermittent electricity is connected to local demand. It will require a great deal of time and a great deal of money to upgrade such a massive system.

An improved grid is critical to making an intermittent plant's economic case. A smart grid will permit additions to the demand load when generation is high due to windy and sunny conditions and reduce the demand load when generation is low due to windless and cloudy conditions.

Dispatchable plants are improved in the exact manner as intermittent plants—by more equitably sharing the relatively small amount of unused electricity produced. Because this amount can be shared more easily, grid operators can choose the best plant to increase or decrease power, based on its efficiency characteristics. Therefore, dispatchable plants can run closer to their peak efficiency.

NATURAL GAS PAIRING

Low-carbon intermittent energy farms are often paired with natural gas power plants to work around the deficiencies of intermittent energy and lack of storage. This results in higher emissions than a fully low-carbon plant, deviating from the original purpose.

Natural gas turbines can be spun up quickly, thus providing on-demand power generation when the intermittent energy source is inactive. The addition of natural gas provides dispatchable capacity to an intermittent capacity resource, avoiding the costs of construction of redundant inter-mittent plants, installation of new storage, or grid improvements.

Natural gas also emits high levels of greenhouse gases from combustion. Having lower emissions than coal does not make natural gas low-carbon! At installations with a large proportion of low-carbon resource intermit-tency, the greenhouse gas from natural gas pairing may be higher than if it were simply a natural gas plant with no low-carbon component, because gas turbines are only efficient at the high temperatures achieved after they have operated for some time.[17] It is certain that the natural gas must be piped in and stored for use, and those pipes and tanks lead to greenhouse gas emissions from methane leaks. Whether or not they are in active use, those natural gas pipes and tanks also tend to kill people through explo-sions and accidents.

Natural gas pairing has a poor return on assets. This is not a compli-cated calculation. Building a second asset that is not in use when the first asset is online and vice versa lowers the return on assets ratio: The assets have increased, but the returns are the same.

Dealing with intermittency using natural gas pairing brings all the nega-tives of natural gas and destroys the value of low-carbon electricity generation.

ADDRESSING INTERMITTENCY VIA GEOGRAPHICALLY REDUNDANT RESOURCES

Figures 18, 20, and 21 depicting intermittency are from 2019 data provided by the Electric Reliability Council of Texas (ERCOT). ERCOT manages 21,200 MW of wind and 1,500 MW of solar capacity from thousands of wind turbines and solar installations. ERCOT, with over 74,800 km (46,500 miles) of transmission lines, spans the state of Texas —a broad geographic swath of 696,241 km^2 (268,581 square miles). These are large numbers describing the area of installations, the number of installations, and the amount of energy generated. Although the aggregate generation is less volatile than any single contributing component, the aggregate itself is still volatile, because weather systems are both very large and very volatile.

An example of a large weather system common to Texas is a hurricane.[18] The hurricane will push a huge ramp-up in generation as more and more wind turbines turn, and then an enormous ramp-down as many wind turbines are simultaneously shut down to protect them from wind speeds beyond their design limits. The large numbers of wind turbines across the large area of ERCOT are not mitigating intermittency; they are weather-correlated resources that reinforce periods of over- and under-generation. This is addressed at greater length in the appendices.

ADDRESSING INTERMITTENCY VIA ANTICORRELATED RESOURCES

From the reliability viewpoint, even better than the uncorrelated plants of the previous section on intermittency are anticorrelated plants! Grid operators want predictability, so we look for sites that are not affected by wind or sun failures in the other.

Examining multiple small energy farms (choice D from Table 8), we notice that the overall capacity factor remains one-third. Raising the total ability to deliver electricity, through anticorrelated plants in and of itself, ensures choosing plant sites that will be offline when others are online. Anticorrelated plants improve reliability but lower the return on assets[19] because the total system capacity factor remains constant, whereas the total system capital investment in assets is greater. Again, every day is game day for the grid.

However, in the real world, none of these plants will be perfectly anticorrelated. The wind turbines are likely to, on average, produce 1,333 MW, with an average curtailment of 333 MW and with the safety margin that two of the sites do not need to produce. But it is possible that all of the sites produce at their peak capacity on a given day, which would produce 3,000 MW of over-generation. Remember: The electricity generated must go somewhere. Will the wind farms be curtailed, or will storage be provided? Although it is improbable that all ten sites would experience failure due to the weather, it is very possible that only seven sites would be producing at one time, leading to a brownout or blackout; the question becomes a determination of the reliability probabilities. Whatever the deficit, it must be made up with storage or with a dispatchable low-carbon power plant—or, in the worst case, a dispatchable high-carbon natural gas plant. Whether demand is met with storage or another power plant, that resource must also be constructed, further lowering the return on assets.

ADDRESSING INTERMITTENCY VIA MULTIPLE ENERGY RESOURCES

The effect of multiple sourcing on intermittent resources such as wind and solar PV and concentrated solar to improve resiliency is an extension of the preceding anticorrelation argument. Resiliency is stated to occur through the mechanism of "when the Sun isn't shining, the wind is blowing, and vice versa." This statement is rhetoric for speeches, not a real phenomenon. Clearly, the Sun does not shine during the night, but it may be covered by clouds during the day. Although it is true that the wind tends to blow at night, this tendency is not reliable; local geography and weather patterns are much better predictors. The peak demand loads are, of course, winter heating and summer cooling.[20]

Let us examine three scenarios, using rounded capacity factors, to provide 2,000 MW through anticorrelated siting.

	Anticorrelated scenario A	Anticorrelated scenario B	Anticorrelated scenario C
	6 wind farms	8 solar PV farms	3 wind + 4 solar PV farms
Nameplate capacity/site	1,000 MW	1,000 MW	1,000 MW
Nameplate capacity	6,000 MW	8,000 MW	7,000 MW
Average capacity factor	33.3%	25.0%	28.6%
Average generation	2,000 MW	2,000 MW	2,000 MW

TABLE 10. Multiple resource reliability scenarios.

Mixing and matching anticorrelated intermittent energy resources changes the system capacity factor but does not change the fundamental insight. It is the anticorrelation that provides resiliency, not the disparate technologies. In other words, disparate energy resources are more likely to be anticorrelated, but that is coincidental to our search for anticorrelation to mitigate intermittency in plant siting.

ADDRESSING INTERMITTENCY VIA DISPATCHABLE STORAGE

We can address intermittency by providing storage, but we still need more power plants because of these two facts: First, the energy for storage must be generated, increasing the demand load; and second, the current demand load does not pause while storage is being filled.

Storage shifts the time of energy use; it doesn't improve how much energy is harvested. Although some of our storage is being charged with what would otherwise be curtailed energy, the total energy required is still larger than the sum of the energy being consumed and the energy being curtailed. More energy farms will need to be constructed to charge storage. The upside of storage is that the cost is probably less than building a full set of mostly idle anticorrelated plants.

EFFICIENCY AFFECTS NEITHER
CAPACITY FACTOR NOR INTERMITTENCY

Efficiency is the ratio of the ability to provide power to a theoretical ability under optimum conditions. Engineers pursue efficiency improvements as they raise the return on assets and lower the amount of construction materials, construction time, and maintenance effort. Efficiency improvements lower costs.

Raising efficiency does not improve generation! Let us consider the case of solar PV. Raising the efficiency of the solar panel does not affect the presence or absence of clouds; intermittency will remain exactly the same. New technological innovation in efficiency does not change the sunrise, the sunset, or the duration of time sunlight falls on the solar panel; the capacity factor will remain the same. When the angle of the Sun, weather, and similar factors are included, the capacity factor will remain essentially 25 percent for the typical solar PV farm.[21]

The efficiency improvement in the solar panel will generate more energy and therefore raise the return on assets of the solar farm. The raised return is realized either by lowering the construction cost for a given solar farm size or by increasing the solar farm size for the same amount of money. The efficiency improvement in the solar panel improves economics, not the intermittency or the capacity factor. A better solar panel will not make the Sun shine more brightly for longer periods of time.

Similarly, raising a wind turbine's height improves the efficiency of a single turbine by lengthening the blades and reaching higher into the windy part of the atmosphere. However, larger turbines need to be placed farther apart to avoid energy losses from turbulence. The collective wind farm has fewer towers and experiences the same weather as before. The useful range of wind remains between 3.5 and 25 m/s approximately a third of the time; therefore, although the cost of installation is reduced, neither the capacity factor nor the intermittency changes. Efficiency improvements in the wind turbine result in cheaper turbines but do not result in more steady and stable wind at the maximum speed of the turbine.

Hydro, geothermal, and nuclear power provide dispatchable examples. One can see yearly capacity factors improve at individual plants as staff gain experience and operate the plant more effectively. The staff is actively directing the machinery of electricity generation. This learning effect is much smaller at wind, solar PV, and concentrated solar farms as the operations depend on passive acceptance of weather and time conditions. However, it is not the efficiency of the turbines or the rest of the plant that affect the capacity factor; it is the learning effect.

Higher efficiency does not result in a higher capacity factor or lower intermittency. Hydropower, geothermal, nuclear, wind, solar PV, concentrated solar—each of these technologies' capacity factors will remain roughly where they are currently, irrespective of underlying efficiency improvements. Likewise, wind, solar PV, and concentrated solar have inherent intermittency that is not affected by efficiency improvements. Siting and prevailing weather of individual wind, solar PV, or concentrated solar farms are the primary determinants of the capacity factor and of intermittency.

ELECTRICITY BILLS

Electricity is used as power; power is the proper amount of energy at the proper time. When the customer turns on a lamp, they want power to illuminate the lamp immediately. This is useable electricity—not too much, not too little, at the desired time—it is power on demand. This simple observation shows us that it is electricity use, power, that drives the electricity bill.

Even the unit of the unit price is telling. The charge is not made against the standard unit of energy (joules). Instead, the utility charges in units of kilowatt-hours, the standard unit of power (watts) multiplied by the time in use (hours).

WHO PICKS UP THE BILL FOR INTERMITTENCY?

The grid has to match generation to consumption, and therefore, intermittent energy must be converted into dispatchable power. A customer turning on a lamp does not want to wait for an intermittent resource to kick in. If we want to use intermittent energy, we must spend money to mitigate the matching of consumption to generation and conversion from energy to power.

If we try to solve this issue with storage, then it has to be purchased, installed, and maintained. If we try to solve this issue by smartening up the grid, then new transmission and distribution networks must be purchased and installed, old networks must be refurbished and upgraded, meters and appliances must be replaced, and control software must be designed, tested, deployed, and supported. Someone has to pay for the storage and smart grids that make intermittent energy possible. There is no free lunch. Raising the amount of money may be difficult, but finance can solve this particular class of technical problems. However, it does create a political problem: Who pays?[22]

Politicians and their appointees on the public utilities commission are keenly aware of the customer and their expectation that lamps light up when turned on. Addressing climate change through low-carbon genera-tion is a secondary concern. However, climate change cannot be ignored. Therefore, we will have to construct a great deal of low-carbon generation capacity, some of which will be intermittent.

SOME QUESTIONS TO THINK ON:

> What is the price of wind generation when the wind is not blowing?
> What is the price of solar PV generation at midnight?
> The price of solar PV at peak generation at noon is very low—in many instances cheaper than coal. What is relevant to determining solar PV's value: average price, marginal price at noon, the spot price at midnight, or the spot price at the moment?
> The price of solar PV at peak generation at noon may even be negative. Does this imply solar PV energy is not just worthless but has negative value?

Now that we speak the lingo of electricity—dispatchable power, intermittent energy, capacity factor, efficiency, storage, smart grid, demand shifting, redundant assets, and curtailment, we will explore scale and the various flavors of low-carbon electricity generation.

CHAPTER 5

TITANIC SCALE

THE WORLD USED 23,696,000,000 MWh in 2018. How much electricity is that? It is a lot. Do not bother to convert this to something like how many 100 W light bulbs could be lit with that amount of electricity. Such comparisons are meaningless and uninformative. Basically, it is so much juice that neither you nor I nor anyone on this planet can comprehend it. The points to remember are that 23,696,000,000 MWh of energy—2,703,000 MW of power—is a lot of electricity.

Humanity will require ever more power,[1] and only entire societies—not any individual—will have an impact on how it is generated. Climate change drives a societal goal of low-carbon electricity generation. What technologies are available to us? What model can we use to value these technologies and which should we pursue and build? What policy tools do we have at hand to guide our choices?

THE NUMBER OF IDEAL LOW-CARBON POWER PLANTS

We may be able to comprehend this effort if we examine how many gigantic 1,000 MW low-carbon power plants are needed to meet the current global electricity demand. These are utility-scale plants, the kind that

governments and large corporations can build, not the type you might set
up in your backyard. Each plant is presumed to have an infinite lifetime
extended by maintenance or replacement.

The 1,000 MW power plant is a good average size for a plant; most are
smaller, a few are larger, but 1,000 MW simplifies the arithmetic. Converting units, the number of ideal 1,000 MW electricity plants needed is 2,703.

	Ideal plants	Nuclear	Geothermal	Hydro	Wind	Solar PV	Concentrated solar
Capacity factor	100%	93.5%	74.0%	43.1%	34.8%	24.5%	21.8%
Number of plants	2,703	2,891	3,653	6,913	7,767	11,033	12,399

TABLE 11. The number of 1,000 MW low-carbon power plants needed by technology type.

We see how big a difference the capacity factor makes. Using a high
capacity factor technology such as nuclear power, we need only 2,900
plants; with a low capacity factor technology such as concentrated solar,
we need 12,400 plants. The number of wind and solar PV plants are con-
servative estimates because these calculations ignore intermittency and
assume that generation is constant.

Geothermal and hydro plants have a significant hurdle to overcome:
plant siting. The geothermal industry is planning to expand three times
over the next fifteen years. However, we would need an expansion by a
factor of over 300 times in the next thirty years to meet global electricity
demand from geothermal alone. Hydropower has a similar problem; most
of the best riverine large-scale sites are taken, and even a small riverine dam
can have a tremendous negative environmental impact. It is unlikely that
either of these technologies can expand as they would need to.

Of course, the real-world solution will include all of these technologies.
We examine each as the only resource available to enable us to compare the
technologies and find valuable insights. For example, we have learned that
geothermal and hydropower would need extreme growth to be considered

at the scales needed. Hundreds or even thousands of such plants would be needed to expand beyond our current generation capacities. A single new hydro plant may be quite large, but how many can we realistically site? Likewise, we realize immediately that aggregating smaller-scale geothermal and hydro plants requires growth of thousands or tens of thousands of new plants.

STORAGE IS NOT AT TITANIC SCALE

To improve the smart grid—in particular using demand load shifting—we need Atlas-size storage. Storage is not a true power source; it does not generate electricity. Storage permits intermittent energy to be stored and released as dispatchable power. Storage always incurs an efficiency penalty to store energy and to release power.

The global amount of electric energy storage is approximately 9,000,000 MWh.[2] That is about 0.038 percent of annual global electricity generation.[3] The aggregate scale of storage is very small.

Pumped hydro is the most common type, contributing about 97 percent of storage. Chemical batteries contribute another 1.4 percent. Compressed air contributes 1.2 percent. Everything else (flywheels, thermal, etc.) makes up the balance of 0.4 percent.

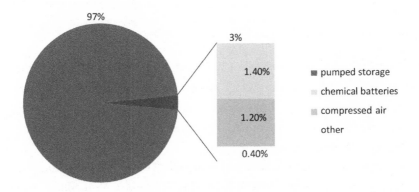

FIGURE 22. Difficult-to-site pumped hydro dominates storage.

We can investigate the scale of individual installations by examining the world's largest storage installation, the Bath County Pumped Storage Station in the US state of Virginia, which has a 24,000 MWh energy capacity, drainable at 3,003 MW peak power, with 79 percent efficiency, in half a day. Of course, the storage may be drained and filled on a daily basis,[4] using 2.5 ideal power plants of 1,000 MW to provide the electricity to fill it in a half day. But if we need this amount of power on a daily basis, why not just build three power plants, tripling the power provided over the full twenty-four-hour cycle to 2,500 MW? To illustrate this more easily, we will create a scenario in which an ideal Bath County[5] is 3,000 MW, 100 percent efficient, with draining and filling time each taking twelve hours.

Plant	Timescale	Energy over day	Peak power	Annual energy
Ideal Bath County	24-hour cycle	0 MWh	3,000 MW	0 MWh
	12-hour discharge	36,000 MWh	3,000 MW	13,149,000 MWh
	12-hour charge	-36,000 MWh	-3,000 MW	-13,149,000 MWh
Three ideal power plants	24-hour production	72,000 MWh	3,000 MW	26,298,000 MWh

TABLE 12. Why build storage when we could build power plants?

What we learn from Table 12 is that storage and generation are fundamentally different. The average power of a dispatchable power plant is essentially peak power.[6] The average power of storage is zero, and an ideal 100 percent efficient system with multiple cycles of discharge and charge storage nets out to zero.[7]

Three dispatchable power plants provide 3,000 MW the entire time. Storage provides 3,000 MW for half the time, and then *consumes* 3,000 MW! Energy-wise, even a huge battery is small compared to a peak-capacity equivalent power plant. For example, annualizing the 36,000 MWh daily output of storage is only 4.1 MW of annualized power. Alternatively, the energy output of a small power plant is equivalent to that of a huge battery.

The largest storage facilities in the world by each type are listed in the following table. Although these storage plants are large, none of them are Atlas-size, although the Bath County Pumped Storage Station has an impressive peak power and endurance. These facilities are physically large. For example, the Bath County Pumped Storage Station occupies an area of 333 hectares (823 acres) and the Hornsdale Power Reserve occupies 20 hectares (49 acres).

Installation	Type	Energy rating	Hours of peak power	Peak power	Annualized power
Bath County Pumped Storage Station	Pumped hydro	24,000 MWh	11	3003 MW	2.74 MW
McIntosh CAES Plant	Compressed air	2,860 MWh	26	110 MW	0.33 MW
Solana Generating Station	Molten salt	1,500 MWh	6	280 MW	0.17 MW
Dalian VFB	Chemical battery, vanadium	800 MWh	4	200 MW	0.09 MW
Project brine4power	Chemical battery, salt water	720 MWh	120	5.83 MW	0.08 MW
Hornsdale Power Reserve	Chemical battery, lithium	185 MWh	1.85	100 MW	0.02 MW

TABLE 13. The world's largest storage in class as of 2019.

The main point is that even the largest storage facility in the world does not have seasonal energy storage capabilities. The nearby city of Washington, DC, with a population of 700,000 people, uses approximately 30,000 MWh of energy daily. Bath County can almost but not quite power this medium-size city through a day—perhaps on a Sunday, when the full demand load is not present. Come Monday morning, the Bath County storage is completely empty, the demand load will increase, and will increase even more so as the Bath County station itself needs to be replenished.

The Hornsdale Power Reserve illustrates that the value of current grid-level storage isn't bulk energy storage but is instead power management. Seventy MW of the 100 MW peak power is reserved for stabilizing the grid over ten-minute periods. This was demonstrated in December 2017 when a power station blew its circuit breakers and went offline. Before another power station's spinning reserve replaced the offline plant, the grid required 7.3 MW of stabilization over a period of eight seconds from Hornsdale. During the time domain of a few seconds, in the power management role, Hornsdale Power Reserve has great value. However, as a long-term (hours to days) or seasonal (days, weeks, months) storage solution for a low-carbon power grid, Hornsdale Power Reserve is quite impractical.

Storage is becoming cheaper and cheaper as technology improves its efficiency, manufacture, and installation. Reducing the marginal cost improves the ability to use the various technologies to provide ancillary grid services, such as voltage and frequency control. Ancillary grid services are the primary use found for current utility-scale storage, as is illustrated by the Hornsdale story.[8] Energy storage will improve the grid, improve the efficiency of electricity generation, and improve demand load shifting.

However, this cost improvement does not improve the ability to use these technologies for bulk storage over long periods of time—for example, matching summer solar energy with winter power use. Storing electricity at the utility scale over the daily cycle, let alone a multiday cycle, is very difficult, given the small (in proportion to a daily cycle) storage that presently exists or is under construction.

Let us investigate storage growth by considering the scale of supporting Belgian electricity consumption over the course of one day. With a population of 11.4 million people, Belgium may be thought of as a small country, a large province, or a very large city. Belgium consumes approximately 243,000 MWh per day.[9]

Number units		Storage type	Unit energy rating
10	Bath County Pumped Storage Stations	pumped hydro	24,000 MWh
85	McIntosh CAES Plants	compressed air	2,860 MWh
162	Solana Generating Stations	molten salt	1,680 MWh
303	Dalian VFBs	vanadium-flow battery	800 MWh
337	Project brine4powers	brine battery	720 MWh
1,312	Hornsdale Power Reserves	lithium-ion battery	185 MWh

TABLE 14. Number of duplicates of the world's largest storage installations by type needed to support Belgium for one day.

Now consider that for intermittent technologies, bulk energy storage may be needed for days at a time rather than just a single day. Neighbors, such as the Netherlands, experience the same societal demand loads and weather conditions at the same time. Belgium will not be able to share; the storage is inflexibly dedicated.

The bad news for scale is that the International Energy Association estimates that storage will grow only about 30 percent over the next few years.[10] An example of this slow growth is the plans of the US state of Arizona to add 3,000 MW of storage by 2030, essentially taking ten years to build a second Bath County. Arizona is undertaking a significantly large project, but not the titanic progress that each and every jurisdiction requires. Storage will remain very small scale.

Existing storage plants are not Atlas-size, future storage plants are not Atlas-size, and their aggregate build-out is not Atlas-size either.

CARBON CAPTURE AND STORAGE IS NOT AT TITANIC SCALE

Carbon capture and storage (CCS) is not at titanic scale in commercial production. As of the time of this book being written, there are only forty-three major commercial facilities utilizing CCS in the world.

CCS is not a power source, but rather a method to remove and store greenhouse gas emissions from coal or natural gas plants. CCS works by simply pulling carbon dioxide out of the smokestack—or preferably from a fuel cell—and then pumping the carbon dioxide deep underground.

The main drawback of CCS is that it always requires more energy than simple coal or natural gas power plants, because it must expend some energy to capture and pump the carbon dioxide. CCS will *always* be less efficient than simple coal or natural gas. Therefore, CCS will never be economically tenable without major policy intervention in the form of carbon taxes or direct subsidies.[11]

CCS technology is not discussed much further because it is not a generation technology, it has not been demonstrated at scale, and it requires immense political intervention to make it economic. Unless these conditions change, it is much simpler to utilize other low-carbon generation technologies.

CARBON-NEUTRAL FUELS AND TRANSPORT

Carbon-neutral fuels are not considered here as a major electricity generation source. The role of carbon-neutral fuels is much better suited to reducing transportation greenhouse gas emissions.

Long cycle-time biofuels require time periods of months and years to achieve sustainability.[12] These fuels are not considered by some governments to be carbon neutral.[13] They are typically grown on farms and require large geographic areas. Biomass crops that take a season to grow can be burned in a few minutes. The following are long cycle-time biofuels:

- Biosynthetic fuels: synthesized by bacteria or algae
- Cellulosic fuels: production of alcohols and diesels from nonfood crops
- Fermented fuels: production of alcohols from food crops
- Biomass direct combustion: combustion of wood and other plants
- Biomass gasification: gasification of wood and other plants

Short-cycle-time carbon-neutral fuels can achieve sustainability in periods of hours or days. These fuels are synthesized in factories with small footprints but require energy inputs. These synthesized fuels also burn in minutes, but they can be regenerated quickly in a matter of hours. Only short-cycle-time fuels are considered further, because they can be produced directly from low-carbon power plants. The following are short-cycle-time carbon-neutral chemical fuels:

- Hydrogen: production of atmospheric hydrogen from water
- Carbon-zero ammonia fuel:[14] production of ammonia from atmospheric hydrogen and nitrogen
- Carbon-based carbon-neutral fuels: production of methane, alkanes, alcohols, and diesels from atmospheric hydrogen and carbon dioxide

Using chemical fuels to generate electricity is a waste. Transport may be partially electrified with electric trains, electric buses, and battery-based automobiles. However, chemical fuels are the densest storage available, with a power density much higher than chemical batteries. It is difficult to imagine battery-powered airplanes as economical in comparison. It is far simpler in many cases—airplanes, heavy vehicles, and smaller ships—to use carbon-neutral fuels. Therefore, carbon-neutral fuels should be reserved for transport.

CHAPTER 6

HYDROPOWER

HYDROPOWER PLANTS ARE THE biggest single generation source power plants humanity has constructed. However, they are difficult to site, and they usually cause great ecological damage.

BOUNDARY HYDROELECTRIC PROJECT

Representative of the large hydropower generation station class, we examine the Boundary Hydroelectric Project, which generates 1,003 MW of power.[1] This power plant produces 3,587,000 MWh of energy annually. It is sited in the Columbia River watershed of the US state of Washington. The dam is quite large—103 m (340 feet) tall, 225 m (740 feet) at its longest length, and 9.75 m (32 feet) thick at its base. The reservoir contains 11,700,000 tons of water, covering 675 hectares (1,668 acres).

The power plant is much bigger! A hydroelectric plant without a reservoir is just a big concrete wall. The total size of the power plant is 65,000 km² (25,200 square miles), the area necessary to catch the rainfall to fill a 11,700,000 ton reservoir. If the area was smaller, say 64,000 km², then 1,000 km² of rain would fall to another dam or would be "lost" in ground water or some undammed river. Although the land area may be shared with other uses—farming, nature preserves, wind turbines, etc.—the land cannot be shared with another hydro plant.

FIGURE 23. Boundary Dam Project, Washington, US.

What would it take to generate all of Belgium's electricity on hydropower, assuming a plant similar to the Boundary Hydroelectric Project? First, it would be incredibly difficult to site a dam in the flatlands of Belgium. Second, France, Germany, Luxembourg, and the Netherlands would have to provide the catchment area, because one Boundary-size plant is over twice the size of Belgium. Last, twenty-five additional such hydroelectric power plants would be needed to power that small country.

RIVERINE HYDROPOWER CANNOT PROVIDE TITANIC-SCALE POWER

As of 2016, the world had 217 riverine hydroelectric power stations 1,000 MW or larger, including thirty under construction. Hydroelectric plants are sited to take advantage of the largest height differences on the largest rivers and to avoid ecological side effects. The best sites are easy to spot and have been built on first. Each successive site is less and less ideal. There simply are not many promising sites left on the planet for big dams.

What about small dams? The Columbia River watershed, meaning the river and all its tributaries, already has 163 dams, of which 103 provide power. As we can see in Table 15, the most common dam provides no power at all. Counting only power-providing dams, the typical riverine hydroelectric plant provides a tiny nameplate capacity of 15 MW. The average hydroelectric plant provides about 400 MW nameplate capacity, but only because the scales are tipped by the very large hydropower stations such as the Grand Coulee Dam at 7,000 MW nameplate capacity.

Most dams serve other purposes, such as providing flood control, improving river traffic navigation, or supplying irrigation water. Power generation is only one of many competing objectives in building dams and allocating the water of the reservoirs.

	All dams	Hydroelectric only
Number of hydro plants	163	103
Total nameplate power	41,057 MW	41,057 MW
Mean nameplate power (average)	250 MW	395 MW
Mode nameplate power (most common)	0 MW	15 MW

TABLE 15. Survey of dams along the Columbia River watershed.

We have calculated previously that we need at least 6,913 large riverine hydro plants to provide for global electricity generation. Subtracting our existing 217 large installations, given that the planet does not have 6,696 more such large sites, we need some 17,000 average 400 MW sites or 460,000 typical 15 MW sites. Even before taking into account the potential ecological disaster of each new hydro plant, we can quickly see that expanding hydropower is not a realistic or practical strategy in anything other than an incremental sense.

You might believe that we can retrofit our many nonpower dams with turbines, even if small ones. Conducting such a survey is worthwhile, but it will not turn up many candidates. During the dam's design phase, the power generation issue has already been evaluated and found lacking, probably due to technical reasons such as an insufficient height difference, insufficient water because of irrigation commitments, or ecological concerns such as fish-spawning routes. It is extremely unlikely that new, more efficient turbine technology is going to have much effect on a reevaluation of power generation.

Run-of-river plants are more extensible.[2] Even so, by its very nature, run-of-river power is small scale, on the order of 50 MW or less of nameplate capacity. We would need a mere 138,000 installations at that size.

MARINE HYDROPOWER CANNOT PROVIDE TITANIC-SCALE POWER

Tidal and wave hydropower cannot provide titanic scale electricity. Marine hydro has low energy densities and lower capacity factors; it comes nowhere close to titanic scale. Tides and currents are well known and predictable, but they are spread across large volumes. Tidal and wave generation will remain small and restricted to very particular geographies.

Mackay[3] estimates 37,500 MW of nameplate power for tidal and wave energy using the waters around the United Kingdom. That is not titanic scale; a single riverine hydro plant, the Grand Inga Dam being constructed in the Democratic Republic of the Congo, will provide 39,000 MW of nameplate power, although admittedly this is the largest riverine hydro plant in the world. Mackay does not estimate the ecological disaster that would accompany creating these tidal energy plants. However, should humanity separate the ocean from the coast, I hazard the ecological impact to be tremendous and horrifyingly negative. As it is, the damage of riverine hydropower is a salutary warning to remain circumspect with tidal and wave energy plants.

Last, the solution is not extensible. The UK is an island, with coastlines on all sides, subject to deep tides and high-amplitude waves. Tidal or wave solutions do not work in land-locked Switzerland or in countries with mild tides and low-amplitude waves, such as Estonia.[4]

PUMPED HYDRO STORAGE IS NOT RIVERINE HYDROPOWER

To create pumped hydro, we need to construct a reservoir at both the top and the bottom of the turbine. Pumped hydro should not be connected to a watershed that can overfill it. Pumped hydro has the single purpose of storage and should not be used for irrigation, navigation, or flood control.

A standard riverine hydro plant is difficult to use as pumped hydro. During a flood, storage is impossible, because the plant's manager is desperately trying to release water by shifting it downstream. During a drought, storage is impossible, because the downstream river, by definition, has no water to pump upstream and store. When the reservoir is at the optimum water level, the turbines are humming along nicely; using that energy to pump upstream the water that just went downstream is pointless.

Strengths	Weaknesses

• Lifetime low-carbon emissions of 24 gCO_2/kWh
• Dispatchable power
• Run-of-river expandable
• Operational costs low
	• Low capacity factor of 39.1 percent
	  ◦ Seasonal dependency
	  ◦ Susceptible to flood and drought
	• Low power density of 0.02 W/m^2
	• Large ecological impact
	  ◦ Often requires human resettlement
	• Few suitable sites left on planet for large-scale riverine hydropower
	• Reservoir is not always single use
	  ◦ Flood control
	  ◦ Agricultural irrigation
	  ◦ Navigation
	• Intermittent (although tidal is predictable)
	• Largest single power plant accident of any type[5]
	• Capital-intensive
	  ◦ Large overnight costs
	  ◦ Large finance charges due to time-to-complete

Hydropower is a great, low-carbon, dispatchable power technology! It will expand in run-of-river projects, new riverine sites in Africa and South America, and turbine upgrades to existing plants. Unfortunately, it is not expandable at the titanic scale required.

CHAPTER 7

GEOTHERMAL POWER

ATLAS WILL ALMOST notice geothermal power's contribution, but if he does, he won't be impressed by it.

HELLISHEIDI GEOTHERMAL POWER STATION

One of the largest of this class, we examine the Hellisheidi Geothermal Power Station, which produces 303 MW of electric power and an additional 133 MW of district heat in Iceland, on the Hengill volcano system.[1] The station has fifty wells, the deepest of which is 2,200 m (7,218 feet).

FIGURE 24. Hellisheidi Geothermal Power Station, Iceland.

What would it take to generate all of Belgium's electricity on geothermal power, assuming a plant similar to the Hellisheidi Geothermal Power Station? First, the stable geology of Belgium is significantly different from the volcanic geology of Iceland. Assuming we could discover such unlikely geothermal resources, we would need thirty-one such plants.

GEOTHERMAL CANNOT PROVIDE TITANIC-SCALE POWER

As of 2016, there were 13,300 MW of geothermal nameplate capacity. The average plant is about 20 MW in size. The current estimates are that the global energy market could expand from 20,000 MW to 32,000 MW over the next fifteen years.[2] Contrast this tiny global geothermal capacity with a single hydropower plant, again in comparison to the Grand Inga Dam, which will provide 39,000 MW, it is easy to see how small the geothermal sector is. To meet global electricity demand, we would need to construct, operate, and manage 205,800 average-size 20 MW geothermal plants.

Should the world invest more in geothermal? No. However, certain geographies should: Japan, Chile, the Philippines, California, Iceland, Hawaii, and New Zealand are all good candidates. All of these are places where geothermal activity is high enough to make geothermal power plants practical. Perhaps the technology may be extended to mine fires, such as the Centralia mine fire in Pennsylvania. These power plants are small but usefully dispatchable.

Strengths	Weaknesses
• Lifetime low-carbon emissions of 38 gCO_2/kWh • Dispatchable power • Second highest capacity factor of 74.4 percent • Operational costs low	• Low power density of 0.054 W/m^2 • Very few suitable sites on planet for large scale geothermal • Highest radiation emissions • Capital-intensive ◦ Large overnight costs

CHAPTER 8

NUCLEAR POWER

NUCLEAR PLANTS ARE DISPATCHABLE, have a high capacity factor, can be built in many locations, and are big—big enough to make a difference. Their main disadvantage is that nuclear plants are extremely capital intensive to build. Their second disadvantage is their perception problem. There are several overblown pearls of conventional wisdom: Nuclear power is too dangerous, its waste is a problem, and it takes too long to construct. These concerns are addressable and, in many cases, already solved.

CENTRALE NUCLÉAIRE DE CATTENOM

Representative of this class, we examine the Centrale Nucléaire de Cattenom with four reactors of 1,300 MW power each, producing 36,739,000 MWh total annually, located on the Moselle River, near Metz.[1] The plant is physically quite small, at 415 hectares (1,025 acres), showing the high power density of nuclear, at 1,253 W/m^2. The capacity factor is 73.8 percent, low for a nuclear plant, due to its operation, which follows the demand load. These four reactors produce about 7.7 percent of French electricity.[2]

FIGURE 25. Centrale Nucléaire de Cattenom.

What would it take to generate all of Belgium's electricity using new nuclear power stations, assuming a similar plant as Cattenom? We would need 2.4 Gen II/III Cattenoms on 1,000 hectares (2,471 acres), 0.03 percent of Belgium's land area. If using a newer design, we would need fifteen Gen IV NuScale-type designs on 540 hectares (1,334 acres).[3]

NUCLEAR CAN PROVIDE TITANIC-SCALE POWER

As we see in Table 16, today's average nuclear reactor is large, about 1,000 MW. There are a mere 451 reactors providing 2,636,000 MWh low-carbon power, about 11 percent of the world's total electricity generation.[4]

	In operation	Under construction	Planned
Number reactors	451	55	81
Total nameplate power	397,000 MW	56,600 MW	97,000 MW
Average nameplate power	880 MW	1,030 MW	1,200 MW

TABLE 16. World nuclear plant inventory.[5]

If the world finished the fifty-five nuclear plants under construction and built the eighty-one planned, adding another 153,600 MW, we would add another 5 percent toward low-carbon electricity. It is easy to see that nuclear power can provide the titanic scale we need.

Nuclear technologies are divided into Gen II/III (pressurized water reactors and boiling water reactors operating between 70 and 140 atmospheres pressure and temperatures around 300°C) and Gen IV (typically, waterless designs operating at low pressures of one atmosphere and temperatures greater than 600°C). See the appendices for a more thorough description.

GEN II/III IS EXPENSIVE

The capital cost of a new Gen II/III plant[6] represents the majority of the total cost of generation, typically as much as 70 percent. These older designs have the following disadvantages:

- On-site construction reduces quality, reduces learning, and reduces the deployment rate.

- High-pressure operation increases the plants' size and cost.

- Low temperatures result in low Carnot efficiency[7] electricity and low-grade waste heat.

- Coolant water can dissociate into explosive hydrogen.

Construction costs range from $2,000/kWh (China) to $9,000/kWh (France), with a global average of $5,000/kWh. However, the main component of the overnight costs is the on-site construction.[8] The construction costs of these older designs are large and coupled with a long time before operation, result in high finance costs.

Construction category	Percentage of the cost
Buildings: excavation, foundation, construction, waste heat cooling–lots of concrete!	47.7%
Nuclear island: reactor and coolant piping–lots of welding!	15.8%
Overhead: site engineering and management–lots of site-specific modifications!	15.3%
Fees, permits, taxes, and commissioning	14.1%
Power island: turbine and generator	7.1%

TABLE 17. Gen II/III overnight cost breakdown.[9]

The simplest method of reducing the overnight cost is to reduce the amount spent on buildings. In order to do this, the buildings have to be smaller in physical size and simpler to construct and must incorporate few site-specific modifications. The reuse of existing waste heat cooling (reusing coal plant cooling towers and river or lake access) may also reduce the buildings component. New markets for low-temperature waste heat—district heat, sewage treatment—can reduce the need to construct waste heat cooling.

GEN IV IS QUICKER AND CHEAPER

Gen IV is attractive because of its low-pressure operation, high temperatures, and quick construction. Gen IV reactors use high-temperature, low-pressure coolants such as molten salt, which simplifies their design and reduces the plants' size. In turn, a simple design and smaller plant leads to faster construction and lowers capital and financing costs. The interesting Gen IV reactors have the following characteristics:

· Factory-based production reduces on-site construction.

· Low-pressure operation reduces the plant size and cost.

- High temperature permits high Carnot efficiency electricity generation.
- High temperature permits multiple process heat applications.
- Inert coolants, such as eutectics[10] of molten salt[11] or lead-bismuth, do not burn, degrade, disassociate, or explode.

FIGURE 26. Inert salt has the consistency and viscosity of water when melted.

The Gen IV plants reduce the main component of the overnight costs by replacing on-site construction with off-site factory manufacturing. Low-pressure operation voids the need for high-pressure containment, permitting reactors to have small physical sizes.[12] Small size allows factory manufacture and shipment of whole components. Even some of the concrete can be prefabricated modules. The lack of a massive pressure containment dome reduces everything else in physical size proportionately to approximately one-eighth the size of Gen II/III plants. Off-site manufacturing with on-site assembly results in shorter times before operation, thus lowering finance costs.

Construction category	Gen III costs	Percentage reduction	Gen IV costs
Buildings: not as much material as Gen II/III	47.7%	75%	11.9%
Nuclear island: not as much welding as Gen II/III	15.8%	50%	7.9%
Overhead: site engineering and management	15.3%	0%	15.3%
Fees, permits, taxes, and commissioning	14.1%	0%	14.1%
Power island: turbine and generator	7.1%	0%	7.1%
Gen IV cost compared to Gen III	–	–	56.3%

TABLE 18. Gen IV estimated 56.3 percent cheaper than Gen II/III.

The smaller buildings of Gen IV are simpler to construct and may be placed underground for both greater protection and a more appealing aesthetic.

GEN IV PROVIDES PROCESS HEAT

Industrial processes, such as cement manufacture, ammonia production, and hydrogen generation, currently have a high carbon dioxide by-product. Gen IV reactors permit decarbonization of these sectors by providing high-temperature process heat.

The molten salts or lead-bismuth eutectics of Gen IV reactor coolants operate in the 600–1,600°C range, perfect for these industrial process heat applications. In contrast, Gen II/III reactors and concentrated solar operate at much lower temperatures of about 300°C and the 250–600°C range, respectively.

NUCLEAR POWER'S HIDDEN SUPER STRENGTH

The attractiveness of nuclear power is clear: It is low carbon, is dispatchable, has no intermittency, has a high capacity factor, is extremely safe, produces negligible waste, and can be built almost anywhere. The real power of nuclear plants is that they are coal plant killers and greenhouse gas assassins. Where nuclear is built, coal plants disappear. Ontario, Canada, demonstrates this phenomenon well.[13]

Technology	2003 generation	2014 generation	Assassination effectiveness
Nuclear power	42%	60%	75%
Coal power	25%	0%	–
Hydropower	23%	24%	4%
Natural gas power	11%	9%	-8%
Non-hydro renewables	0%	7%	29%

TABLE 19. Ontario stopped burning coal in 2014.

The reverse case is also true. Where nuclear plants disappear, fossil fuel plants appear. Germany and Japan both shut down nuclear plants after the Fukushima Daiichi accident and restarted or built new coal and natural gas plants. A prime example is the high-carbon 1,654 MW Moorburg CHP Plant, sited in Hamburg, Germany, which initially opened in 2015 as nuclear plants were shut down.

FEAR OF NUCLEAR POWER

Unlike other forms of electricity generation, nuclear power has the unique problem of public misunderstanding and fear to overcome. As fear is an irrational, negative emotion, it is naturally difficult to address with facts and figures.

So, first of all, let me assert my firm belief that the only thing we have to fear is fear itself—nameless, unreasoning, unjustified terror which paralyzes needed efforts to convert retreat into advance.

—FRANKLIN DELANO ROOSEVELT, INAUGURAL ADDRESS, MARCH 4, 1933

NUCLEAR WASTE

All of a nuclear plant's high-level waste is collected and stored. The global amount from the past sixty-five years is a very small 371,000 tons.[14] Over a third of a million tons of high-level waste sounds like a lot, but we need some perspective. Volumetrically, all of the world's high-level waste from commercial nuclear power could fit in New York City's Yankee Stadium with lots of room to spare. The waste would pose no threat to the surrounding neighborhood for several decades, and it would take decades to fill the stadium completely. Of course, Yankee fans would rather see their baseball team than cement cylinders, and so Yankee Stadium is not an ideal repository. The point is that we obtain titanic amounts of electricity for this Lilliputian amount of safely contained waste.

The Moorburg coal plant produces 371,000 tons of greenhouse gas every sixteen days. That waste is not collected. That waste is not stored. That waste is full of particulates that affect our lungs, full of mercury and lead that affect our brains, and full of radioactive elements. That waste is vented directly into the atmosphere. That waste causes asthma and other diseases. That waste kills people each and every day.

	371,000 tons spent fuel	371,000 tons greenhouse gas	371,000 tons greenhouse gas
Source	Global nuclear industry	Moorburg coal plant[15]	Pastoria energy facility[16]
Number of plants	over 623 reactors[17]	1 coal plant	1 natural gas plant
Creation period	65 years	16 days	83 days
Pollution	collected, stored, monitored	uncaptured, unstored, vented to biosphere	uncaptured, unstored, vented to biosphere
Generation value	10% of global electricity for six decades	2% of German electricity	2% of Californian electricity

TABLE 20. Comparison of entire global nuclear industry to a single German coal or Californian natural gas plant.

Unlike greenhouse gas emissions, nuclear waste is collected and stored in casks designed to last decades. Gen II/III waste is a long-term issue that society has several decades to solve. Even if the issue is not solved permanently, at worst the casks can simply be replaced near the end of their service life. Nuclear waste is not a short-term issue in which the waste is vented to the atmosphere, causing climate change today.

Good news! Society already has a solution for high-level nuclear waste. We simply separate the spent nuclear fuel into short-lived radioactive fission products (isotopes of krypton, barium, cesium, iodine, etc.) and long-lived radioactive actinides (isotopes of uranium, plutonium, etc.). We place the long-lived actinides where they belong, inside a Gen IV reactor. Gen IV reactors use fuel more effectively, and so the actinides undergo fission to produce short-lived fission products. The separated fission products decay to background radiation levels after only 300 years. Society has storage solutions for short-lived 300-year fission product waste. The fission products amount to only a few percent of the spent nuclear fuel, both by mass and by volume. Refer to the appendices for an extended explanation.

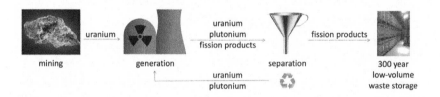

FIGURE 27. Gen IV spent nuclear fuel recycling and waste. Uranium, plutonium, and other actinides are simply returned to the reactor.

The waste of nuclear electricity generation is not a major problem.

EXPLOSIONS AND ACCIDENTS

First, nuclear power plants are incapable of a nuclear explosion. They cannot produce a mushroom cloud and cannot flatten a city. At worst, a nuclear reactor can reach very high temperatures, melting the fuel and reactor vessel into ooze. As the ooze melts the reactor vessel, more ooze is created, lowering the fuel density.[18] As the fuel becomes less dense, the chain reaction stops, the temperature goes lower, and the ooze freezes in place.

Unfortunately, Gen II/III nuclear plants contain water as a moderator and as a coolant to carry the heat from the reactor to the turbine. When water gets hot very quickly, either it forms into superheated steam or it dissociates into hydrogen and oxygen gases. The Chernobyl accident was a steam explosion, the force of which blew the reactor's top off. The heat then started a fire that burned for days, creating an updraft. It was the steam explosion and the fire's updraft that ejected radioactive particles into the environment. The Fukushima Daiichi accident essentially comes down to hydrogen explosions. The heat of the uncooled reactors dissociated water into hydrogen and oxygen. The hydrogen combusted and blew the reactor buildings apart. It was these hydrogen explosions that allowed radioactive particles into the environment.

Gen IV plants do not contain water. They use molten salt or molten metals to carry the heat from the reactor to the generator. Without water, it is impossible to create steam or dissociate water into hydrogen. No water means zero chance of a steam explosion. No water means zero chance of dissociation into explosive hydrogen gas. Molten salts or molten metal coolants are effectively pre-oozed and therefore dilute quickly. In an accident, the fuel oozes directly into the pre-oozed coolant, trapping the fuel and the fission products for later cleanup. See the appendices for an extended description.

RADIATION AND CANCER

Yes, nuclear plants produce a lot of radiation. However, all of that is inside the reactor. Coal, natural gas, and geothermal plants, during normal operations, output more radiation to their surroundings than do nuclear power plants.

Epidemiological analysis of the Chernobyl accident indicates that, in total, fifteen have died from radiation-induced thyroid cancer, a few thousand have been diagnosed with thyroid cancer, and some 4,000 to 11,000 extra people will have radiation-induced cancer over their lifetimes. The appendix goes into greater detail.

In contrast, while shutting down its low-carbon pollution-free nuclear plants, Germany continues to burn coal, killing 4,350 people[19] each year due to cancer and respiratory diseases. Forty-six hundred of the deaths from German coal are attributable from the phase-out of nuclear power since 2011, projected to be an additional 16,000 deaths by 2035.[20] At the global level, burning fossil fuels kill millions. Coal, petroleum, and natural gas mining—let alone burning—result in 3,000 to 4,000 deaths each year.

The analysis of Fukushima Daiichi indicates that lifetime increased cancer risk due to the accident for Fukushima residents is approximately 1 percent. We can see how small this is in comparison to a far more common activity our society permits—smoking tobacco.

Lifetime increased cancer risks from Fukushima Daiichii	Lifetime increased cancer risks from smoking
• 1 percent for residents	• 22 percent for male smokers • 12 percent for female smokers • 1 percent for nonsmokers via secondhand smoke

Contracting cancer is a serious health concern, but we must have a sense of proportion. The tragedy of the Chernobyl accident is less relevant than the tragedy of continuing to burn coal or pump natural gas. The induced cancers of the Fukushima Daiichi accident are insignificant compared to the induced cancers of smoking tobacco. The long-term global health effects of burning of coal, natural gas, and other fossil fuels are immediate and far more dangerous than an elevated risk of cancer from trace radiation emissions that only occur in a serious accident. See the appendices for an extended discussion of radiation and other issues presented by nuclear power.

Strengths	Weaknesses

• Second lowest lifetime emissions of 12 gCO_2/kWh
• Dispatchable power
• Highest capacity factor of 93.5 percent
• Highest power density of 1,253 W/m^2
• Large-scale plants practical to build
• Small waste profile
  ◦ 100 percent of waste captured and stored
  ◦ Waste is compact and low-volume
• Operational costs low
	• Highest fear and misunderstanding factor
	• Gen II/III specific problems
	  ◦ Requires siting near cooling water
	  ◦ Overnight costs highest
	  ◦ Finance costs high due to time-to-complete
	• Gen IV specific problems
	  ◦ Regulatory uncertainty

Nuclear is a great, low-carbon, dispatchable power technology that can deliver both electricity and process heat. It has broad siting flexibility and is expandable to the titanic scale needed. The issue of cost is being addressed with Gen IV plants. The issue of unpopularity due to fear is difficult to address.

CHAPTER 9

WIND ENERGY

WIND FARMS HAVE BOTH low power density and very low energy-collection density. Although the wind tower footprint is relatively small, at a few hundred square meters, dual-use activities such as agriculture are possible. The aggregate wind farm catchment area is quite large. This is similar in concept to the catchment area of a hydro plant.

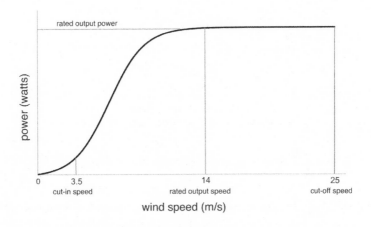

FIGURE 28. Capacity factor: Wind is only useful between 3.5 and 25 m/s for a typical turbine.

Although hydro plants have a lower energy-collection density than wind, they collect the energy in one place, requiring a single large connection point and transmission system. Wind turbines collect energy in multiple turbines, effectively requiring each tower to have its own connection point and transmission system, even if the towers on the farm share a backbone transmission cable.[1]

Wind is both variable in strength and variable as to whether it is blowing or not. Wind may blow above the design rating[2] for a wind turbine (25 m/s, 56 miles/hour), above the efficient speed of the turbine (20 m/s, 45 miles/hour), at the efficient speed (14 m/s, 31 miles/hour), below the efficient speed (10 m/s, 22 miles/hour), at the cut-in speed (3.5 m/s, 7 miles/hour), or not at all.

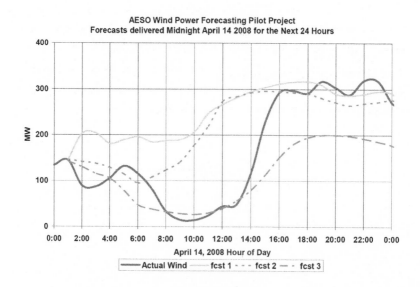

FIGURE 29. Wind prediction models have limited success in forecasting intermittency.

Predicting the wind is helpful for running the grid. Generation must always match demand load, but the actual intermittency is what determines how much power the grid operators can call on. Improving prediction doesn't improve wind, but it does reduce the strain intermittent wind puts on dispatchable power plants and the grid.

In Figure 29, the startling aspect is not the divergence of actual from the prediction but the fact that the actual wind ramp rate is so steep over such a limited period of time.

Offshore wind can achieve lower intermittency and higher capacity factors. However, offshore wind is also more expensive and dangerous to operate.

ROSCOE WIND FARM

Representative of this class, we examine the Roscoe Wind Farm.[3] It is located in Texas, the US state with the most wind farms. Nameplate capacity is 782 MW of power, producing approximately 2,174,000 MWh of energy annually. The Roscoe Wind Farm is quite large, with 627 towers spanning 400 km² (154 square miles) across four counties. The land is otherwise used for agriculture.

What would it take to generate all of Belgium's electricity on wind energy, assuming weather and turbines similar to those at the Roscoe Wind Farm and assuming that intermittency will be zero? It would take 41 Roscoes of 25,700 wind turbines spanning a multiuse catchment of 16,400 km² (6,332 square miles), about 53 percent of Belgium's land area.

GANSU WIND FARMS PROJECT

The Gansu Wind Farms Project[4] is a megaproject incorporating many smaller wind farms located in the province of Gansu, China. Together, they currently supply 8,000 MW using over 7,000 towers on thousands of square kilometers, approaching the same area as the US state of Rhode Island. Once completed, the projected nameplate capacity will be 22,000 MW and therefore representative of the largest wind farm project in the world as of 2020. This is comparable to the 22,500 MW of the single, dispatchable hydroelectric power station of the Three Gorges Dam.

This project illustrates the problems associated with any megaproject.[5] The wind farm sits in the Gobi Desert, 1,600 km (994 miles) away from the major population centers of China, and requires new transmission infrastructure to deliver the energy. As an illustration of the seriousness of this problem, 39 percent of the energy available in 2015 was not delivered, because there was no long-distance transmission or local consumer for the electricity. Similar to other multibillion-dollar projects, a decade of building is required, starting in 2009 and scheduled to be completed in 2020.

WIND CAN PROVIDE TITANIC SCALE ENERGY

The world has enough space to build onshore or offshore wind, at three or four times the assets needed to overcome the low capacity factor limitation. Wind typically requires rezoning changes in terms of scenery and noise and rezoning changes for transmission lines.

WIND NEEDS STORAGE TO PROVIDE TITANIC SCALE POWER

Wind is a low-grade energy source that requires storage and a smart grid capable of dealing with its intermittency to operate at titanic scale power. Assuming that it is even achievable to build the amount of storage needed, determining which stakeholders should pay for the massive amounts of new storage and major improvements to the grid to make wind energy a practical power source is of interest to taxpayers, ratepayers, financiers, and wind farm owners.

WIND TURBINE EFFICIENCY

Raising the height and size of wind turbines raises the total efficiency of single turbines. However, the efficiency of an entire wind farm improves only slightly with taller and larger turbines, because the larger towers are spaced farther apart to avoid air turbulence from the other turbines. Air turbulence decreases a turbine's efficiency. Essentially, the efficiency of larger turbines is cancelled out by the smaller number of turbines spread further apart.

Improving the efficiency of windmill blades and turbines reduces cost and results in cheaper wind energy. Turbine efficiency does not affect the fundamentals of capacity factor and intermittency. Those are driven by the weather, not the technology. A lack of wind cannot be overcome by efficiency improvements. Better economics does not result in better physics.

Strengths	Weaknesses

• Lowest lifetime emissions of 11 gCO2/kWh
• Large scale plants practical to build
• Small waste profile
• Operational costs low
	• Intermittent energy
	  ◦ Requires short-term storage
	  ◦ Requires long-term (10-100 hours) storage
	  ◦ Requires seasonal storage
	  ◦ Requires smart grid
	• Low capacity factor of 34.8 percent
	  ◦ Peaks at night when demand is low
	  ◦ Seasonally peaks in spring, during lowest demand
	  ◦ Requires integration of excess plant
	• Low power density of 3-4 W/m²
	  ◦ Extremely large farms
	  ◦ Often requires new transmission infrastructure
	• Falls and slips resulting in injury or death is one of the most common industrial accidents, a consideration for 150-200 m (500-650 feet) towers
	• Capital-intensive
	  ◦ Large overnight costs

Wind is a low-carbon, intermittent technology. However, its intermittency can be mitigated with storage and smart grids.

CHAPTER 10

SOLAR PV ENERGY

SOLAR PV WORKS BY exposing photovoltaic panels to sunlight. Essentially, sunlight knocks the electrons of the panel into a higher energy state; this creates a charge in the electric circuit.

Utility-scale solar PV plants have the most misunderstood cost structure of the low-carbon plants. This is due in part to the conflation of utility-scale solar PV and residential-rooftop solar PV.

It is true that the price of a photovoltaic panel has dropped like a rock, but the panel is only a part of the cost for a utility-scale plant. We also must account for the purchase or lease cost of the extensive area the plant uses. The shadow beneath the plant is mostly useless.[1] No sunlight means no farming or grazing, so there is little dual use of the land. Moving panels from land to water merely moves the shadow, reducing aquatic growth. PV panels contain water-soluble heavy metals and can leak if they are damaged. PV panels need frequent cleaning of dust to maintain efficiency, requiring water resources. Replacement costs will be frequent: The best panels last about twenty-five years. Lower-quality panels need replacement in less than ten years.

The low cost of solar PV often detracts from the major disadvantage—the panels do not work during the night—loss of generation during half the day results in a low return on assets. Solar PV needs storage or supplementary generation to meet nighttime demand.

TOPAZ SOLAR FARM

Representative of this class, we examine the Topaz Solar Farm, in the US state of California.[2] Its nameplate capacity is 550 MW of power, and it produces 1,282,000 MWh of energy annually, with a 26.6 percent capacity factor. The site spans 19 km^2 (7 square miles) and took approximately three years to build.

FIGURE 30. Topaz Solar Farm.

What would it take to generate all of Belgium's electricity on solar PV energy, assuming weather and infrastructure similar to those in the Topaz Solar Farm and assuming that storage exists for nighttime power and that intermittency will be zero? It would take sixty-nine Topaz farms extending over single-use areas of 1,311 km^2 (506 square miles), which is about 4 percent of Belgium's land area. Assuming that Belgium receives the same radiance as California or that the required storage can be built are each as big a stretch as assuming zero intermittency.

SOLAR PV CAN PROVIDE TITANIC SCALE ENERGY

The world has enough space to build solar PV at four or five times the capacity needed to overcome its low capacity factor limitation.

SOLAR PV NEEDS STORAGE TO PROVIDE TITANIC SCALE POWER

As with wind, solar PV is a low-grade energy source requiring storage and a smart grid capable of dealing with its intermittency to operate at titanic scale power. Although peak solar energy partially coincides with peak human activity, the key word in the sentence is *partially*, and so storage is needed to provide power at other times and to smooth out intermittency. As can be seen from Figure 31, intermittency has a strong impact on minute-to-minute reliability.

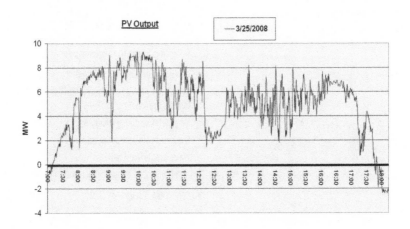

FIGURE 31. Solar PV is very dependent on cloud cover.[3]

Again, assuming that it is even achievable to build the amount of storage needed, determining which stakeholders should pay for massive amounts of new storage and major improvements to the grid to make solar PV energy practicable is of interest to taxpayers, ratepayers, financiers, and solar PV farm owners.

SOLAR PV EFFICIENCY

Improving the efficiency of solar panels reduces their cost and results in cheaper solar PV. However, the efficiency increase does not affect the fundamentals of capacity factor and intermittency. Those are driven by the planet's rotation, the site's latitude, and the local weather, not technology. A lack of sunlight cannot be overcome by efficiency improvements. Better economics does not result in better physics.

Strengths	Weaknesses

• Lifetime low-carbon emissions of 44 gCO_2/kWh
• Peaks during daylight, roughly matching peak demand
  ◦ Seasonally peaks in summer, a high demand time
• Large scale plants practical to build
• Operational costs low
	• Intermittent energy
	  ◦ Requires short-term storage
	  ◦ Requires long-term (10–100 hours) storage
	  ◦ Requires seasonal storage
	  ◦ Requires smart grid
	• Low capacity factor of 24.5 percent
	  ◦ Seasonally low in winter, a high demand time
	  ◦ Requires integration of excess plant
	• Low power density of 5–20 W/m^2
	• Installability low as weather pattern and latitude dependent
	• No generation during the night
	• Competes with residential solar PV
	• Waste profile uncertain
	  ◦ Currently solar PV is difficult to fully recycle
	• Capital-intensive
	  ◦ Large overnight costs

CONCENTRATED SOLAR ENERGY

CONCENTRATED SOLAR FARMS focus the Sun's light using mirrors to heat the plant's coolant. The coolant is then used in a turbine to generate electricity. The coolant is typically molten salt, which can retain the heat as a thermal reserve for several hours.[1]

Concentrated solar with its accompanying storage may be thought of as dispatchable in an hourly time frame. As long as the plant's thermal reserve of coolant holds out, the plant can respond to the demand load. In general, we cannot predict whether tomorrow will have sufficient sunlight to ensure dispatch over a daily time frame. Enlarging the size of the storage reservoir's capacity does not help, because extra coolant without extra heat does not make extra energy. To take advantage of an enlarged reservoir, we must either enlarge the plant to receive more solar energy or receive outside energy to match the enlargement.

If we enlarge the plant to ensure a two-day thermal reservoir of coolant, rather than one day, we have extra assets that only earn a return if the second day is cloudy. Alternatively, we could use all assets on sunny days and shut down a coal plant. Of course, this means that, on cloudy days, all our assets are idle because our generation drops to zero, and the grid will bring the coal plant back online. Enlarging the plant changes the capacity, not the capacity factor.

SOLANA GENERATING STATION

Representative of this class, we examine the Solana Generating Station, in the US state of Arizona.[2] Its nameplate capacity is 250 MW of power, and it produces 776,000 MWh of energy annually, with a much higher than the average 35.4 percent capacity factor. The site spans 7.8 km² (3 square miles) and took approximately three years to build. The plant is able to store heat for six hours.

FIGURE 32. Solana Generating Station.

The primary coolant for Solana is an oil eutectic. This then heats the secondary coolant, molten salt, which provides storage and heat for steam turbine power generation.

What would it take to generate all of Belgium's electricity on concentrated solar energy, assuming weather and infrastructure similar to those of the Solana Generation Station, assuming storage exists for nighttime power, and assuming intermittency will be zero? It would require 112 Solana farms extending over single-use areas of 874 km² (337 square miles), which is about 2.9 percent of Belgium's land area. But assumptions of a vast twelve-hour storage capacity, California sun, and zero intermittency are ludicrously optimistic.

CONCENTRATED SOLAR CAN PROVIDE TITANIC SCALE POWER

The world has enough space to build concentrated solar at five or six times the capacity needed to overcome the intermittency limitation. It does have the advantage that peak solar energy nearly coincides with peak human activity; the thermal reservoir can extend that time coincidence.

Unlike wind or solar PV, concentrated solar farms do not need improvement in short-term storage, but they do need improvements in long-term (ten to one hundred hours) and seasonal storage. Short-term intermittency is dealt with through its own short-term coolant storage. However, concentrated solar is still a low-grade energy source that requires Atlas-size storage to operate titanic-scale power to deal with the lack of dispatchability over multiple days of bad weather. And again, determining which stakeholders should pay for the massive amounts of new storage and major improvements to the grid is of interest to taxpayers, ratepayers, financiers, and concentrated solar farm owners.

CONCENTRATED SOLAR EFFICIENCY

Improving the efficiency of mirrors, heat exchangers, and coolants reduces the cost and results in cheaper concentrated solar farms. However, the efficiency increase does not affect the fundamentals of capacity factor and intermittency. Those are driven by the planet's rotation, the site's latitude, and the site's weather, not technology. A lack of sunlight cannot be overcome by efficiency improvements. Better economics does not result in better physics.

CONCENTRATED SOLAR PROCESS HEAT IS NOT TITANIC-SCALE

Like nuclear plants, concentrated solar uses a relatively hot coolant, in the 250–600°C range. This is on the cool side for industrial processes but still useful. Unfortunately, concentrated solar farms need to be sited where the sunlight is. It is unlikely that the conditions for a concentrated solar farm are present where the existing factories and chemical works are sited. Remote deserts are the current favored choice of most such farms. Factories and chemical works would have to be built new, next to a concentrated solar farm.

The second issue with concentrated solar process heat is the inherent intermittency of solar energy. In particular, chemical works are designed for 24/7 throughput. Some reactions take on the order of days. Intermittent heat that stops in the middle of the reaction or factory process is not tenable. Again, backing up these processes with natural gas lowers the return on assets and voids the value of heat generated from low-carbon technologies. This issue is reexplored in the appendices.

Strengths	Weaknesses

• Lifetime low-carbon emissions of 27 gCO_2/kWh
• Peaks during daylight, roughly matching peak demand
  ◦ Seasonally peaks in summer, a high demand time
• Large scale plants practical to build
• Short-term dispatchable power
  ◦ Includes own short-term thermal storage on the order of hours
• Operational costs low
	• Intermittent energy
	  ◦ Requires long-term (10-100 hours) storage
	  ◦ Requires seasonal storage
	  ◦ Thermal storage on the order of days or weeks is difficult to extend economically
	• Low capacity factor of 21.8 percent
	  ◦ Seasonally low in winter, a high demand time
	  ◦ Requires integration of excess plant
	• Low power density of 15 W/m^2
	• Installability low as weather pattern- and latitude-dependent
	• Capital-intensive
	  ◦ Large overnight costs

CHAPTER 12

BELOW UTILITY-SCALE LOW-CARBON CHOICES

NONE OF THESE building-size applications are titanic scale. However, at the scale of individual or company investment, they are important complements to energy conservation. Approaches such as district heating are large scale but are utility-level actions, not within residential or commercial scope. Installing or upgrading any individual residence is effectively beneath measurement at the utility scale. Commercial energy use is a bright spot, because many commercial buildings use large amounts of electricity, peak usage is typically during the day, and they often have large roofs for solar PV or concentrated solar.

Nonetheless, with intelligent attention to building codes, tax rebates, clever technologies, carbon taxes, along with conservation efforts, micro-generation can, in aggregate, reduce the demand load on utilities.

HEAT PUMPS

Heat pumps are just fantastic! The installation of heat pumps is easily one of the top methods for the typical building to save total energy and reduce carbon emissions. The reason is that 10 to 18 percent of the building's total energy (and over 50 percent of residential usage) is used to heat buildings. Heat pumps work in a variety of geographies and climates. If you own a building, you should investigate the suitability of installing a heat pump for it.

The pumps do not truly generate energy, but a small amount of electric power may be applied to the pump to move a large amount of heat between the inside and outside of a building. The most common heat pump is the refrigerator. The second most common is the air conditioner, a refrigerator put together backward.

A basic description of heat pumps is included in the appendices.

MICRO SOLAR PV

Micro solar PV is essentially building-size solar PV, and it suffers from the same issues as utility solar PV—storage and intermittency. Utility-scale solar PV can use long-distance transmission to avoid local issues, but geographic location is a huge determinant of micro solar PV value: It is great in sunny Mexico, terrible in dark Finland.

Micro solar PV is usually a fixed installation on a roof; it rarely employs the sun-tracking mechanisms common to utility-scale solar PV. The generation efficiencies are often determined by roof geometries. The return on assets is usually quite low because redundant panels are oriented to optimize generation at given times of the day. Panels oriented southward produce the most energy, but panels oriented westward produce the most value because the peak electricity demand is in the evening.

Without storage, the building must use the electricity immediately. Commercial buildings are most suitable for solar PV, because the large surface area of commercial roofs, windows, and parking lots, coupled with the largely daytime demand corresponding to their daytime generation, makes these attractive venues for solar PV installation. Commercial firms are structured to make solar PV investment and depreciation easier to finance. With storage, the building can store the energy for later use as power, but the extra cost of storage installation lowers the return on assets. Commercial installations avoid this extra cost due to the immediate use of solar energy.

Residential buildings are mostly empty during the day, when the energy is generated, and so, in general, they are not well suited for solar PV. Likewise, industrial buildings need much larger amounts of electric power and process heat than can be provided by the relatively small surface area: The energy collected from a factory's roof and parking lot is marginal with respect to the total energy that the factory will use.

NET METERING

Net metering is the concept of sending a building's generated electricity to the grid when its solar PV over-generates. When the building's solar PV under-generates, electricity is taken from the grid. If the solar PV is sized correctly, they should net out in terms of energy. Unfortunately, some believe—and, in many jurisdictions, it is law—that the costs also net out.

This false argument is that the grid is like a bank, where you can put electricity in only to withdraw it later. We must remember that electricity is consumed as it is produced and that overproduction cannot be consumed. The value at the time of generation is not the same as the value at the time of consumption.[1] The grid is not like a bank.

To change the timing of electricity consumption, we need storage. If the building has its own storage, then the question of who pays for it is clear: The building's owner does; they bought the storage. If the grid is providing utility-scale storage, then storage is more like a physical storage space, where one pays a fee to access and use the space.

Who pays for microgeneration? There are two models that society can adopt: The first option is called the *free-rider* approach, wherein micro-generators sell electricity at the same retail price that they pay. These users do not pay for storage. They do not pay for the convenience of being able to use grid electricity. They do not pay for grid maintenance, even though they need access to sell electricity. They receive retail price, a preferential premium that other generators do not. And they receive a premium for selling less valuable small-scale intermittent energy rather than the more valuable large-scale dispatchable power.

The second option is the *fair-price* approach, in which the microgenerators sell electricity at wholesale and pay grid maintenance fees. They do pay for storage and for the convenience of being able to use grid electricity. They do pay for grid maintenance, because they need access to sell electricity. They receive wholesale price, not a premium. And they are charged for the overhead of integrating small-scale intermittent energy.

Microgeneration can be a useful—if rather small-scale—contribution to a low-carbon electricity future. However, society needs utilities to invest in titanic-scale power plants, not subsidize free riders.

MICRO CONCENTRATED SOLAR

Micro concentrated solar, which uses sunlight to heat a coolant, is essentially building-size concentrated solar. Residential concentrated solar has some big pluses. Although geographic location is a huge determinant of concentrated solar value, this is offset by four attributes: Concentrated solar produces heat directly, and heating is over half the energy use in residential buildings. Concentrated solar inherently has a heat reservoir and so can be stored for use in the evening, the time of major energy use in residential buildings. Concentrated solar can share coolant, pumps, and its heat reservoir with a heat pump system, making both together cheaper than each in isolation.[2] Atmospheric-pressure water is an ideal coolant in a residential installation because it is readily available from city water systems, and because it exactly matches human comfort temperature ranges, it can have a secondary use as hot water in washing, bathing, and cooking.

MICRO CONCENTRATED SOLAR PROCESS HEAT

Commercial or industrial concentrated solar tends to be less useful in general. Specific application niches, such as laundries or bakeries, may find the process heat from concentrated solar preferable to solar PV.

MICROGENERATION CHOICE VALUATION

In general, individual buildings should first concentrate on conservation—insulation, energy-efficient appliances, and energy reduction if possible. The installation of these solutions then follows this general value pattern, although specific geographies and needs dictate specifics.

	Residential	Commercial	Industrial
Highest value	conservation	conservation	conservation
High value	heat pumps	solar PV with storage	--
Median value	concentrated solar	heat pumps	--
Lowest value	solar PV with storage	concentrated solar	--

TABLE 21. Economic value of microgeneration choices by sector.

MICROGENERATION SCALE VALUATION CHART

Microscale valuation is very different from society's choices for utility-scale power. Often, it is a matter of the home's or company's suitability for the energy choice, investment capability, and desire for a smaller utility bill.

The installation is on a single site and rarely requires the environmental, financial, and safety regulatory studies needed at the utility scale. The kilowatt capacity, at best a few megawatt-hours per year, does not attract the public utility commission's attention. The capacity factor is irrelevant, because the majority of microgeneration sites remain connected to the grid. Again, the energy choice is primarily driven by the site's suitability for the application.

Technology	Dispatchability	Storage required	Smart grid required	Installability
Heat pumps	–	+1	+1	+1
Storage	+1	–	+1	+1
Concentrated solar	+1	–	+1	-1
Solar PV	-1	-1	-1	-1

TABLE 22. Valuation comparison for micro-scale installation.

Putting together a comparison model, we can see that for heat pumps, dispatchability is mostly irrelevant; it is moving heat around, not producing it. Storage of heat is not required, resulting in a plus-one parameter. Heat is not connected to the grid at all, and not needing a smart grid results in a plus-one parameter. Although the installation is somewhat dependent on geography, in general, heat pumps are widely applicable throughout the world, resulting in a plus-one parameter.

Storage is dispatchable without grid reliance—so it does not need a smart grid—and can be installed worldwide, resulting in plus-one parameters across the board.

Concentrated solar has a short-term dispatchable heat reserve for a plus-one parameter, although its thermal reserve is still dependent on sunny weather. Concentrated solar does not depend on the grid for a plus-one parameter.

Solar PV has uniformly minus-one parameters. Solar PV without storage is not dispatchable. Practical solar PV negatively depends on storage, and if exported to the grid, requires a smart grid to take full advantage. Neither concentrated solar nor solar PV is installable in general, because installation negatively depends on latitude, weather patterns, and a building's orientation.

CHAPTER 13

SOCIETAL VALUATION OF UTILITY POWER SOURCES

IN ORDER TO COMPARE the six low-carbon technologies, we need to develop a valuation model. Each significant attribute of the model is represented by a weighted valuation parameter. The weights placed on each parameter may differ between you and me, but the model itself may be shared.

Each parameter has a coefficient of the type listed here:

· Numerical: for example, the known capacity factors
· Positive: all positive values represent an advantage
· Negative: all negative values represent a disadvantage

The comprehensive set of valuation attributes is as follows:

· Greenhouse gas intensity
· Capacity factor
· Storage requirement
· Installability

· Dispatchability

· Excess capacity assets

· Global plant overhead

· Smart grid requirement

· Power density

Our six low-carbon generation technologies are all proven and none is prohibitively expensive. It is interesting to observe that all low-carbon electricity sources have relatively high capital costs and relatively low operational costs.

One attribute notably absent is cost. This is a valuation model, not a pricing or costing model; it is the value that these technologies provide that matters, not their cost. Compare the *Mona Lisa* to a child's drawing. The cost of the *Mona Lisa* is just a few dollars more than the cost of a crayon portrait on your refrigerator, but the valuation is very different.

Mona Lisa
Artist: Leonardo da Vinci
Costs:

• Canvas

• Oil paints

• Labor of a master

Value: Priceless cultural artifact

Untitled
Artist: Maggie Fry
Costs:

• Construction paper

• Crayon

• Labor of a child genius

Value: In the eye of the beholder

Valuation in this context describes how useful a low-carbon electricity generator this technology is compared to the others. Costing is a separate and less interesting exercise. High-value solutions will retain their value despite their cost, and low-value solutions will never be worth the money spent.

TECHNICAL, MONEY-INSOLUBLE PROBLEMS

These nonfinancial attributes each represent a difficult technical problem. We *may* solve this class of problems through applying the brainpower of scientists and engineers; we may also be unable to solve these problems in a timely or complete manner. We are unable to buy a solution, but we are able to fund research and development.

GREENHOUSE GAS INTENSITY

This valuation attribute is a straightforward measure of the amount of carbon dioxide equivalents per kilowatt-hour (gCO_2/kWh) the solution generates. Remembering that the greenhouse gas intensity is the single most important determinant of lowering carbon emissions, large numbers indicate a loss of value; therefore, this attribute's parameter is negative. Less is more!

Wind	Nuclear	Hydro	Concentrated solar	Geothermal	Solar PV
-11	-12	-24	-27	-38	-44

TABLE 23. Greenhouse gas intensity parameters. The units are grams of CO_2 per kWh.[1]

We may expect incremental progress in reducing these greenhouse gas intensity parameters. Hydropower, geothermal, nuclear, wind, solar PV, and concentrated solar plants each have mining, fabrication, steel, cement, and transportation energy embedded in their construction and operation. Until industrial process heat comes from zero-carbon resources and transport fuels are fully carbon-neutral fuels, these emissions will remain.

CAPACITY FACTOR

The capacity factor reflects the average generation ability of a source. This valuation attribute is the capacity of the solution over the course of the year, represented in percentages. High capacity factors have a high-value contribution.

Hydropower, geothermal, nuclear, wind, solar PV, and concentrated solar plants each obey the fundamental physical laws of the universe, and although an individual plant may have better capacity factors, their average capacity factors will not change much. As an example, the Earth will keep rotating and night will come no matter how fantastic our solar plant is.

Nuclear	Geothermal	Hydro	Wind	Solar PV	Concentrated solar
93.5%	74.0%	39.1%	34.8%	24.5%	21.8%

TABLE 24. Capacity factor parameters.

We may expect progress in plant operations only. Operations can be improved, but repair, replacement, and maintenance will still require a minimum amount of time.

STORAGE REQUIREMENT

This valuation attribute is a qualitative measure of the need to mitigate intermittency through storage. Plus-one indicates that storage is not required and effort is avoided. The constraints of long-term and seasonal storage technology make large-scale construction unlikely. Minus-one indicates that the resources require energy storage and extra assets need to be constructed. The parameter's range from minus-one to plus-one reflects the distinction between storage *required* for conversion of inter-mittent energy resources to dispatchable power sources and storage *nice to* have for dispatchable power sources to run at their most efficient capacity factors.

Hydro, geothermal, and nuclear technologies do not require storage; their parameter is plus-one. Wind and solar PV do require storage; their parameters are minus-one. Concentrated solar has inherent thermal storage that allows it to meet demand over the hourly cycle, a day's worth of demand, but loses that capability on the next cloudy day, and so its parameter is zero.

Hydro	Geothermal	Nuclear	Concentrated solar	Wind	Solar PV
+1	+1	+1	0	-1	-1

TABLE 25. Storage requirement parameters.

Whichever technology choices are made, more storage should be built to improve the grid and its ability to balance the demand load.

INSTALLABILITY

Installability is the attribute that measures how easy it is to place a plant just anywhere both now and in the future. The high, glacier-covered mountains of Norway make excellent opportunities for hydropower. The same power station in the flatlands of Belgium would be pointless. This valuation attribute's parameter is qualitatively set from plus-one for nuclear power, for which most sites can be placed, to minus-two for hydropower, which is extremely site-dependent and for which most of the best global locations have already been taken. We may expect improvements in preparing technologies for installation and in the installations themselves.

Nuclear plants are relatively easy to site and have great potential for wider siting and so nuclear is assessed as plus-one. Reducing water use dramatically improves future siting. Computer modeling will herald new turbine efficiencies with high-temperature super-critical carbon dioxide turbines, which will allow waste heat to be cooled with air rather than water. Magnetohydrodynamic systems will enable new waterless turbine technologies altogether. Most promising of all, high-temperature process heat encourages siting of nonelectric, thermal nuclear power next to industrial plants, skipping the intermediate step of electricity generation.[2]

Wind technology has limited installability improvement, related to site selection based on better weather prediction. Technologies to move wind offshore are helpful, but the sea is a harsh environment, and maintenance is much more difficult. Wind is assessed as a zero in this category.

Solar PV and concentrated solar technologies have more room for installability improvement. New siting is possible with solar panels able to tap into broader regions of spectrum. New coolants for concentrated solar will enable broader temperature ranges to be tapped, again leading to new siting opportunities. However, solar will remain dependent on both latitude and the prevailing weather pattern of the site. Because it depends on unpredictable technical innovation and site particulars, solar is assessed as zero in this category.

Geothermal power plants will always remain extremely dependent on the geology of the site and geothermal is assessed as minus-one. Hydropower siting is assessed as minus-two, because of the lack of available sites and additionally incurs ecological damage to the watershed. In the drive to build low-carbon electricity generation, a lack of installability essentially limits the contribution of geothermal and hydropower to existing sites and slow, limited, incremental growth. These technologies remain included for comparison purposes.

Nuclear	Wind	Solar PV	Concentrated solar	Geothermal	Hydro
+1	0	0	0	-1	-2

TABLE 26. Installability parameters.

INTERMITTENCY IS INSOLUBLE

Intermittency simply is, or it isn't. We already know the problem of intermittency is insoluble.

DISPATCHABILITY

This valuation attribute measures the responsiveness of dispatchable power to the ever-changing demand load. Hydropower, geothermal, and nuclear power plants are nonintermittent dispatchable resources; their parameters are plus-one. Concentrated solar farms, which also rely on an intermittent solar energy resource, are only semi-dispatchable; its parameter is zero, because the inherent storage components are dispatchable over the period of hours. Wind and solar PV farms, which rely on an intermittent energy resource, are simply not dispatchable; their parameters are minus-one.

No amount of research will change how much time the Sun shines each day or how much wind is blowing. We will develop better prediction models for weather and electricity consumption demand. However, predictions will not change the intermittency of the energy resource or the dispatchability of the generation plant. These predictions will be used to determine how much dispatchable power is required to be added to intermittent energy to meet demand load.

Hydro	Geothermal	Nuclear	Concentrated solar	Wind	Solar PV
+1	+1	+1	0	-1	-1

TABLE 27. Dispatchability parameters.[3]

We will develop slightly better storage to mitigate intermittent energy, but that will be limited by scale to the second–minute–hour time domains to mitigate short-term intermittency. Monthly or seasonal storage[4] time domains will remain beyond our engineering reach unless some fantastic and incredible scientific breakthrough is made. We will improve the grid to become smart. That improvement will be limited by the magnitude of the risk we are willing to take on and the amount of money we invest in the grid.

NONTECHNICAL, MONEY-SOLUBLE PROBLEMS

These valuation attributes each represent nontechnical problems. Solutions may simply be purchased because no major advances in science or engineering are required. These problems are solved by the liberal application of cash. We do not have to invent a new solution; we merely have to pay for an existing solution. We must determine how to pay for the solution, but society has invented finance and has trained financiers to assemble and leverage money. Determining how to pay is a technical financing issue, not a technical engineering problem.

The main financial problem is the political issue of who pays.

EXCESS CAPACITY ASSETS

This valuation attribute represents excess capacity assets that contribute to a lower return on assets. In order to meet the practical capacity needed from a plant, we take the desired capacity and divide by the capacity factor to arrive at the necessary excess nameplate capacity. This results in a larger plant. Larger plants are more difficult to design and have larger operational and maintenance overhead. Larger plant size can be overcome by overspending money. This valuation attribute's parameter is insensitive as to whether the capacity is built as a single plant or multiple smaller plants.

The attribute's parameter is negative because it is calculated as a deficit from the ideal 1,000 MW plant. This calculation assumes steady-state generation and zero intermittency. Real-world non-zero intermittency for wind, solar PV, and concentrated solar energy farms results in even more negative values.

	Nuclear	Geothermal	Hydro	Wind	Solar PV	Concentrated solar
Capacity factor	93.5%	74.0%	39.1%	34.8%	24.5%	21.8%
Assumed intermittency	–	–	–	0	0	0
Nameplate capacity needed	1,070 MW	1,344 MW	2,558 MW	2,874 MW	4,082 MW	4,587 MW
Excess nameplate capacity	–70 MW	–344 MW	–1,558 MW	–1,874 MW	–3,082 MW	–3,587 MW

TABLE 28. Excess capacity assets parameters.

Hydro and geothermal plants' generation capacity is determined by site selection, not by the plant design. Therefore, this parameter is somewhat unreliable for these technologies.

Low capacity factors raise the size of wind, solar PV, and concentrated solar farms several times, straining siting requirements and increasing complexity. These farms are very big, complex industrial sites. Examples are the large geographic areas over which to coordinate wind turbines or the effort to dust tens of thousands of square meters of solar PV cells or concentrated solar mirrors.

GLOBAL PLANT OVERHEAD

Low capacity factors lead to extra effort needed to address the demand load. This valuation attribute reflects the negative implications of the increased overhead of siting, construction, and operation of the truly vast numbers of low-carbon power plants the world needs. This valuation attribute's parameter is insensitive as to whether the plant is an excess capacity asset.

	Nuclear	Geothermal	Hydropower	Wind	Solar PV	Concentrated solar
Capacity factor	93.5%	74.4%	39.1%	34.8%	24.5%	21.8%
Assumed intermittency	-	-	-	0	0	0
Number of plants	2,891	3,633	6,913	7,767	11,033	12,399
Global plant overhead	-188	-930	-4,210	-5,064	-8,330	-9,696

TABLE 29. Valuation parameters of number of 1,000 MW low-carbon
power plants to meet global demand.

The parameter is calculated as a deficit from the 2,703 ideal 1,000 MW plants needed across the globe, indicating the greater overhead of low-capacity resources. The calculation assumes steady-state generation and zero intermittency. Again, real-world non-zero intermittency for wind, solar PV, and concentrated solar energy farms results in even more negative values.

Nuclear, wind, solar PV, and concentrated solar can be installed in the numbers needed, and so calculating the parameter as above is straightforward.

The ability to construct additional hydro and geothermal plants is determined by site selection and not by the plant design. Therefore, this parameter is somewhat unreliable for these technologies.

SMART GRID REQUIREMENT

Upgrading the grid to faster and more accurate demand and generation balancing—in particular, demand management—will improve the output of all power plants, and so in all scenarios is a prudent, cost-effective, and wise investment choice. We should build a smart grid to improve all generation.

How smart the grid becomes is simply a function of how much money is spent to make it so. The distinction here is between the *required* upgrades to mitigate intermittency and *optional* upgrades to improve all generation.

If the smart grid is not required, then the valuation attribute's parameter is plus-one as effort is avoided. Resources requiring a smart grid result in a rating of minus-one. Constructing the smart grid requires more effort, more maintenance, and more time to have an impact. As low capacity factor and high-intermittency resources are added, these grid improvements must be deployed faster and must be more sophisticated technically.

Hydro, geothermal, and nuclear plants have values of plus-one because they benefit from but do not require improvements to the grid. Concentrated solar has a value of zero because its inherent storage component requires a lower investment in the grid to meet demand over the hourly cycle. Wind and solar PV have values of minus-one because they require improvements to the grid to be economical.

Hydro	Geothermal	Nuclear	Concentrated solar	Wind	Solar PV
+1	+1	+1	0	-1	-1

TABLE 30. Smart grid requirement parameters.

A grid with a high penetration of intermittent sources has high complexity because of transmission to geographically disparate and large-area wind and solar farms, grid reconfiguration due to natural phenomena such as storms, transmission to utility-scale storage, transmission to distributed storage (such as electric vehicles and residential batteries), and demand load shifting.

High complexity usually translates into high costs. Who will pay—ratepayers, taxpayers, all electricity providers equally, or all electricity providers proportional to the value they receive? There is a case to be made that ratepayers, taxpayers, and providers should all pay to improve the grid. Again, all providers benefit from an improved grid, and therefore, their ratepayers benefit. Taxpayers also benefit from a more efficient energy sector, resulting in an improved economy. On the other hand, the only group that *requires* a smart grid is the high-intermittency electric energy providers.

Although grid improvements are addressable with money, addressing the imbalance between demand and generation very quickly reaches high costs and diminishing returns. Building a grid smart enough to deal with high levels of intermittent energy generation will be expensive.

POWER DENSITY

A geographically dense energy source is more cost-efficient: less construction effort, less maintenance, closer proximity to demand centers, lowering transmission costs, and lessening the ecological impact. However, with the application of money, we can purchase land (or sea) to be set aside for electricity generation. A geographically dense energy source is cheaper to convert to useful power. This valuation attribute's parameter is therefore positive because the high energy-density technologies have high value.

The amount of land used to produce energy has been falling over centuries, from the preindustrial biofuels of forests and waste crops through animal muscle and windmills to postindustrial fossil fuels and nuclear power. Renewable energy farming of wind and solar reverses this trend and even extends it to areas of the sea.

Nuclear	Solar PV	Concentrated solar	Wind	Geothermal	Hydro
1,253 W/m^2	5–20 W/m^2	15 W/m^2	3–4 W/m^2	0.054 W/m^2	0.02 W/m^2

TABLE 31. Power density parameters.

Nuclear power requires very little land. The facilities are densely packed into a very small area. A typical nuclear plant has a land plot of around 100 hectares (250 acres) but uses only a few tens of hectares for the actual installation, the remainder being a security zone.

Nuclear plant sizing changes very little with respect to generation output. In this particular case, the difference between a 1,000 MW and a 2,000 MW plant is unlikely to be reflected in the physical area delimited by the site boundary. Wind, solar PV, and concentrated solar farm sizing changes in a linear relation to generation output.

Wind turbines have a small tower footprint, spread out over a very large area, typically hundreds of square kilometers. These areas include access roads and transmission lines. The total area often has dual uses for farming or ranching. However, these structures are not compatible with cities or areas frequented by people; just as with other industrial structures, they are subject to fire, collapse, and other accidents. Because of these hazards, wind turbines are often zoned 500–1,500 m (0.3-1 mile) away from frequent human activity.

> Dual use does not change the energy density of wind or solar. The presence or absence of sheep on a wind farm has little effect on the wind's strength or on turbine siting.

Solar PV and concentrated solar farms have a large land use, over a large area—typically hundreds of hectares. The area is rarely dual use, except rooftop installation, as was discussed in the preceding microscale sections. It is difficult to see utility-scale solar plants as ecologically beneficial to the site they occupy.

Geothermal installations are similar to wind, with a small surface footprint collecting over a large total area not dedicated to power production, leading to dual uses.

Hydropower has the largest land requirement, and typically, the reservoirs needed are not dedicated to power production. The reservoirs often also provide flood control, crop irrigation, and navigable waterways.

VALUATION COMPARISON

Table 32 presents the parameters of each valuation attribute of the model.

Technology	Greenhouse gas intensity (gCO_2/kWh)	Capacity factor	Dispatch-ability	Storage requirement	Excess capacity assets	Global plant overhead	Smart grid require-ment	Power density	Installability
Non-titanic scale possibilities (due to lack of siting possibilities), ranked by greenhouse gas intensity									
hydro	-24 g	39.1%	+1	+1	-1,558 MW	-4,210	+1	0.02 W/m²	-2 (filled)
geothermal	-38 g	74.4%	+1	+1	-344 MW	-930	+1	0.054 W/m²	-1 (specific)
Storage for comparison									
	type-dependent	type-dependent	+1	-	-	-	+1	type-dependent	type-dependent
Titanic-scale possibilities, ranked by greenhouse gas intensity									
wind	-11 g	34.8%	-1	-1	-1,874 MW	-5,064	-1	3-4 W/m²	0
nuclear	-12 g	93.5%	+1	+1	-70 MW	-188	+1	1,253 W/m²	+1
concen-trated solar	-27 g	21.8%	0	0	-3,587 MW	-9,696	0	15 W/m²	0
solar PV	-44 g	24.5%	-1	-1	-3,082 MW	-8,330	-1	5-20 W/m²	0

TABLE 32. Valuation comparison ranked by greenhouse gas intensity.

Note that costs and efficiency (another impact on cost) are not part of this valuation model. To realize the value that a given solution provides, we can set financiers on the problem of purchase. Although it may cost more to purchase an old, musty da Vinci, the higher price presumably reflects a higher value. Alternatively, if you place a higher value on a modern Maggie Fry original, the higher price will be paid there. It is the value, the return, that is important, not the purchase price.

You can now assign the importance of these attributes by choosing your weights, W. As a tangible example, the valuation model of hydropower is written out. To obtain numerical results, determine the weights as a fraction summing to one, and calculate the valuation of each technology. To compare the technologies, rank them according to the valuations.

$$\text{Valuation}_{\text{hydro}} = -24 \cdot W_{\text{GHG Intensity}} + 39.1 \cdot W_{\text{Capacity Factor}} + 1 \cdot W_{\text{Dispatchability}}$$

$$+1 \cdot W_{\text{Storage Not Required}} - 1{,}558 \cdot W_{\text{Excess Capacity Assets}} - 4{,}210 \cdot W_{\text{Global Plant Overhead}}$$

$$+ 1 \cdot W_{\text{Smart Grid Not Required}} + 0.02 \cdot W_{\text{Power Density}} - 2 \cdot W_{\text{Installability}}$$

NON-CONTENDERS

Hydropower and geothermal simply don't have installability. No matter their other attractive features, the world does not have thousands of suitable sites to generate the millions of megawatts needed.

Storage doesn't generate power at all; it merely stores it. However, the comparison of storage to generation technologies is useful in illustrating the broader context.

FOUR CONTENDERS

That leaves four contenders to both compete and collaborate to solve the problem of generating titanic amounts of low-carbon electricity.

Wind energy emits the lowest amount of greenhouse gas among our low-carbon options. It has high intermittency, a low capacity factor, poor density, and low install flexibility. It requires new storage capacity and grid upgrades.

Nuclear power is the second lowest for greenhouse gas emissions. It is dispatchable and has a high capacity factor, high power density, and high install flexibility. It can take full advantage of, but does not require, new storage capacity and grid upgrades.

Concentrated solar energy is a medium greenhouse gas emitter among our low-carbon resources. It has high intermittency but some dispatchability due to intraday storage. It has the lowest capacity factor, poor density, and low install flexibility at the utility scale. It requires long-term storage capacity and grid upgrades.

Solar PV energy is the highest emitter of greenhouse gas among our low-carbon power options. It has high intermittency, a low capacity factor, poor density, low install flexibility at the utility scale, and it is useless at night. It requires new storage capacity and grid upgrades.

NUCLEAR TO COMPETE

We can see that nuclear power compares favorably in almost every category. This is not too surprising. Society's experience has been to move from pre–Industrial Revolution harvesting of natural resources over large geographic areas to post–Industrial Revolution manufacture of products in compact factories. Analogously, nuclear provides cheap electricity generated from a dense energy source in a compact power factory.

Nuclear power is dispatchable, with no intermittency problem. Nuclear plants have high capacity factors. Nuclear plants have a small area footprint, have demonstrated the most flexible installation requirements, and can be sited local to the demand load. Nuclear plants are very safe and have a tiny waste stream. Nuclear power has no significant technical challenges to overcome at the utility scale. A global fleet of nuclear plants has a low overhead of siting, construction, and operation. Nuclear plants benefit from but do not require improvements in storage technology or construction of storage capacity.

Nuclear power plants have few externality costs. Super-sized or multiply redundant assets are not needed. Scientific breakthroughs and engineering innovations in storage are not needed. Extra, complementary storage assets are not needed. Grid improvements that can control the demand load are not needed. Costs such as waste and decommissioning are fully accounted for.

Superficially, nuclear power appears to have a poor financial case based on purchase price, but it has a strong return on assets. The main problem is the financing of such singularly capital-intensive assets that are the on-site-construction, high-pressure Gen II/III designs. This financing problem is addressed by moving to factory-produced low-pressure Gen IV designs with on-site assembly.

WIND, SOLAR PV, AND CONCENTRATED SOLAR TO COLLABORATE

Wind, solar PV, and concentrated solar technologies are possible solutions. The main value of these technologies is their low-carbon footprint. This low-carbon footprint must not blind us to their drawbacks. Wind, solar PV, and concentrated solar compare unfavorably in almost every category. These technologies have significant technical challenges to overcome at the utility scale. They can support a first-world electricity grid up to 20 percent or 30 percent penetration before their lack of storage becomes economically crippling.[5] Installability can be improved but will remain fundamentally centered on local weather patterns, site latitude, and site conditions. A global fleet of energy farms has a high overhead of siting, construction, and operation.

Because these technologies are high intermittency, have a low capacity factor, and are low-density electricity farms, they are severely constrained: They entail a low return on assets due to needing super-sized or multiply-redundant assets. Their feasibility depends on innovation in storage technology—particularly, long-term storage to address seasonal intermittency. Meeting the demand load requires the construction of large amounts of storage capacity. Supporting intermittency requires major improvements to the grid.

These technologies will never overcome their insoluble intermittency problem. They are best suited to niche applications, where their lack of dispatchability is not an issue. At the residential and commercial micro levels, they can usefully support individual buildings with matched consumption and generation.

Financially, these technologies appear to be on a reasonable footing. However, their low return on assets and reliance on externalities show fundamental financial weaknesses at the utility scale.[6]

POWER DENSITY, POLITICS, AND SOCIETY

Some wind and solar advocates actively promote renewable technologies from a political or societal viewpoint. The argument is that distributed and renewable technologies shift wealth—and, therefore, political power—from corporate utilities to consumers, resulting in a more desirable society.

Regardless of the merits or desirability of such a shift of political power, the argument that distributed energy will precipitate political or societal shifts does not hold weight. Utility-operated distributed and renewable technologies are still controlled by utility companies. Utilities are necessary; microgeneration, such as rooftop solar, is simply not enough energy to power a first-world lifestyle—mines, farms, factories, etc. Neither low-density energy sources nor extreme energy conservation will lower significantly the external energy cost of a first-world lifestyle.

It is not utility profits that make the use of fossil fuels undesirable. Coal and natural gas are undesirable because burning these carbon-based fuels results in acid rain, harmful particulates, and greenhouse gas emissions that cause climate change.

Likewise, it is not whether power is generated in a distributed manner or not that affects politics and society; it is how densely power is consumed. Centrally generated power affects the cost of power, not societal norms. High levels of electricity use, through dense, low-cost power, enable societies to be more desirable places in which individual people lead more fulfilling lives.[7]

Urban agriculture, which provides higher-quality foods nearer to people's kitchens, depends on high power consumption.

FIGURE 33. High-power agriculture.

Recycling materials requires a lot of low-cost power to convert trash into treasure.

FIGURE 34. High-power recycling.

Large amounts of cheap electricity enable our society to entertain and educate itself after the Sun sets.

FIGURE 35. High-power entertainment.

ANYTHING BUT NUCLEAR!

If you are in the *anything but nuclear* camp, then you need to examine the following issues.

Storage may partially solve wind's, solar PV's, and concentrated solar's intermittency and low capacity factors. Remember: The test of storage is not whether evening power can be supplied from morning and midday wind and sun energy. The test is whether storage capacity exists to supply power for the evening when there is no morning and midday wind and sun energy, perhaps for several days in succession. First, we need to determine the amount of storage needed to guarantee 100 percent power over seven days a week, twenty-four hours a day, through different seasons of the year and different yearly weather patterns. If that power cannot be guaranteed, will the power be rationed as brownouts for all or blackouts for some? Is it even possible to build that amount of storage? If two generation lulls occur within a short time span of each other, we must predict the time needed for energy production to top up the energy storage. We must determine whether this storage depends on unknown, undiscovered, or undeveloped technologies. We must predict when such technologies will be realized and quantify our confidence in the realization timeline. We must also characterize the value society will place in such a confidence factor. Finally, we must determine a fallback plan if the unknown, undiscovered, or undeveloped technologies fail to be realized and also the plan should they fail to be realized within the given timeline.

Siting pumped hydro is a very different proposition between mountainous Norway and the flat plains of Belgium. The footprint area of giant chemical batteries and underground compressed air is very different in a rural or urban environment. Can flywheels or superconducting capacitors provide the amount of storage capacity needed for a city of 500,000 people?

Investments in smart grid technology may partially solve wind's, solar PV's, and concentrated solar's intermittency and low capacity factors. To do so, we must determine the investment needed for a smart grid (transmission improvements, demand load shifting, penetration of smart appliances, grid topological changes, and variable pricing) capable of guaranteeing 100 percent power over seven days a week, twenty-four hours a day, through different seasons of the year and different yearly weather patterns.

Constructing extra plants may partially solve wind's, solar PV's, and concentrated solar's intermittency and low capacity factors, but this results in a low return on assets, including the assets needed to charge storage. We must determine the amount of duplication needed to guarantee 100 percent power over seven days a week, twenty-four hours a day, through different seasons of the year and different yearly weather patterns. Then we can describe the rationale that permits the assumption that this is enough redundancy. We must characterize the confidence factor in the duplication factor and characterize the value society will place in such a confidence factor. Finally, we must create a fallback plan if the confidence factor is too optimistic.

CONFIDENCE IN NUCLEAR AS A SOLUTION

Nuclear power is so easy that humanity built a thermal reactor in 1942 simply by interleaving graphite and uranium. A fast reactor was constructed in 1951. Everything else in nuclear science and engineering has been an incremental—not breakthrough—discovery. Improvements in nuclear power since those two reactors have been in different cooling options, better materials and construction, longer-lasting equipment, more-efficient turbines, and reduced costs. By 1973, a mere two decades of engineering progress later, several 1,000 MW reactors were built or were under construction. The appendices further examine the problems that conventional wisdom associates with nuclear power—safety, waste, and weapons proliferation.

Nuclear power doesn't require storage or smart grid improvements to guarantee 100 percent power seven days a week, twenty-four hours a day, through different seasons of the year and different yearly weather patterns; nuclear plants operate independently of the weather. Nuclear power is not intermittent; almost all downtime is due to scheduled maintenance. Unlike storage, investment in Gen II/III and Gen IV nuclear does not depend on unknown, undiscovered, or undeveloped technologies. Gen IV nuclear will benefit from research to improve plant life, modify waste profiles, lower maintenance costs, increase build velocity, and reduce capital costs, but these are improvements, not basic feasibility.

Nuclear plants have a good return on assets. We do not need to build duplicative and redundant plant capacity beyond what is needed to cover maintenance periods. With the average nuclear plant's 93.5 percent capacity factor, only one extra plant is needed for every ten built to achieve a system capacity factor of 100 percent.

Seventy years on in the Atomic Age, no fundamental scientific or engineering breakthroughs are needed to utilize nuclear fission power. For the last forty-five years, dispatchable, high capacity factor, low-intermittency, carbon-free, Atlas-size 1,000 MW power plants have been sited, built across multiple geographies, and operated without regard to weather and without damage to the climate.

Society can believe with a high confidence factor that nuclear power provides a workable solution.

SOLVING THE PROBLEMS OF NUCLEAR POWER

All the problems of the nuclear solution are about money, not about technology. We also know that a problem that may be solved with money is not a problem. Let us examine the following three questions.

The first question is how to build nuclear plants at a lower capital cost. The answer is construction that uses low-pressure, factory-fabricated, Gen IV technology and proper financing terms. Nuclear power's financial problem of high capital costs is solved with moving to prefab Gen IV reactor designs.

The second question is what is more important—a lower capital cost or a higher value. The answer is a higher value. France built a fleet of nuclear power plants in the 1970s and 1980s and has some of the lowest greenhouse gas emissions in Europe (25 percent of the average). Germany has shut down nuclear power plants and built a fleet of renewables (nearly 40 percent of electricity generation), but it continues to operate fossil fuel power plants (nearly 40 percent of generation is coal) and is on the wrong side of European greenhouse gas emissions (150 percent over the average). The French capital investment in nuclear may have had a higher initial cost, but that investment clearly has a higher value than the German capital investment in renewables.

The third question is which are easier to change—physical laws of the universe or human laws of society. The answer is human laws. The laws of thermodynamics complicate the generation of dispatchable power out of intermittent energy, creating compact and cheap storage, and building a demand-managed grid. Human law is so much easier to change. King Cnut had no control over the tides, but he had a lot of control over his empire's infrastructure, investment, energy, and tax policies.

Nuclear power is currently too expensive. With a few tweaks of the legal code, its virtues can be recognized, its financing made available, and its construction encouraged.

COMPARING LOW-CARBON GENERATION IN 2050

The United States commissioned a fantastic report from the National Renewables Energy Laboratory (NREL) to determine the feasibility of installing a high proportion of renewables within the United States in the year 2050.[8] The authors' intent was to examine the feasibility of high-penetration scenarios and different mixes of renewable technologies; it was not meant to examine the value of such a system. However, with some back-of-the-envelope calculations, we can make some inferences about the value of different systems by comparing relative emissions.

The bar chart in Figure 36 presents four scenarios: the 2050 Baseline, the NREL 2050 RE-ETI 80%,[9] the 2018 French Mix, and the 2050 Nuclear Mix.

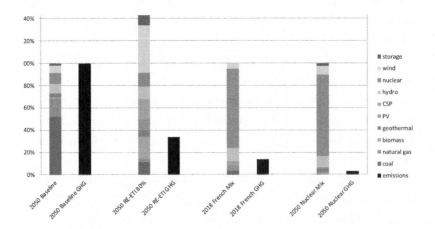

FIGURE 36. Capacity mixes, asset amounts, and greenhouse gas emissions.

The NREL 2050 Baseline is the business-as-usual scenario, in which the 2050 generation proportions are nearly the same as today: We continue to burn a lot of coal to generate electricity. The other scenarios are measured with respect to this scenario's generation and greenhouse gas emissions.

The NREL 2050 RE-ETI 80% scenario represents a future projection of 80 percent penetration of renewable generation, under optimistic assumptions of technology innovation and progress. This scenario shows a clear reduction to about 35 percent in greenhouse gas (GHG) emissions relative to the 2050 Baseline, assumes that society continues to burn coal and natural gas (about 15 percent of generation), and shows a 40 percent increase in nameplate capacity assets compared to the 2050 Baseline for the same electricity generation, resulting in a lower return on assets.

I have added the 2018 French Mix scenario, which applies the current-day French capacity mix proportionately, assuming zero technology innovation over the 2050 Baseline. The emissions in this scenario drop to about 15 percent relative to the 2050 Baseline and under half of the NREL 2050 RE-ETI 80% scenario. Nameplate capacity remains about the same, resulting in a better return on assets than the NREL 2050 RE-ETI 80% scenario.[10]

I have also added the 2050 Nuclear Mix scenario by taking the 2050 Baseline and completely replacing high-carbon coal and natural gas generation with low-carbon nuclear power. This achieves the lowest emissions of all of the scenarios, about 3 percent of the 2050 Baseline, because it has no high-carbon component at all. Nameplate capacity remains about the same, resulting in a better return on assets than the NREL 2050 RE-ETI 80% scenario.[11]

The 2018 French Mix, with 70 percent nuclear, has no risk of revolutionary technology improvements failing to materialize, because it uses existing technology. Our high-carbon economies can focus primarily on lowering emissions rather than on being distracted by the secondary task of grid improvements and storage expansion. The 2050 Nuclear Mix of 80 percent nuclear and 20 percent renewables, with no high-carbon component at all, achieves super-low emissions.

Dispatchable, installable, grid-friendly nuclear power is simply the easiest primary generation choice to construct and to integrate—so easy that the French have already done so.[12] To implement the 2018 French Mix or the 2050 Nuclear Mix scenario across the globe, we do not require optimism. We only require the will and the project management to start construction. We are good to go.

CHAPTER 14

POLICY ACTIONS

GOVERNMENTS CAN ENCOURAGE low-carbon power with economic incentives. Governments can support and encourage low-carbon research, build low-carbon test plants, and subsidize low-carbon commercial pilot plants. Governments can shepherd planning and permitting of commercial low-carbon plant construction. Likewise, governments can discourage high-carbon technologies with economic disincentives—raising taxes, imposing fees, and creating diminishing quotas for high-carbon electricity generation.

PIGOUVIAN TAXES

A Pigouvian tax is a policy lever to make externality costs explicit.[1] Governments can discourage high-carbon technologies by setting taxes and fees on them. Taxing carbon fuels, especially coal, provides a clear signal to individuals and markets. The point of the Pigouvian tax is not to raise revenue but to deter unwanted behavior—in this case, the burning of high-carbon fuels such as coal, natural gas, and oil.

The tax need not be onerous, but it must be credible and comprehensive. As the market finds substitutes, such as low-carbon electricity, the tax should gradually rise. By setting the tax on the carbon content of fuels, it does discriminate against coal, one of the dirtiest fuels, in favor of natural gas. This is acceptable because coal is twice as dirty as natural gas. However, the tax also discriminates against both high-carbon coal and high-carbon natural gas in favor of the low-carbon technologies of hydropower, geo-thermal, nuclear, wind, solar PV, and concentrated solar.

PIGOUVIAN CARBON CAPTURE AND STORAGE

A Pigouvian carbon tax can change the calculation that a fossil fuel plant with carbon capture and storage (CCS) capability is always more expen-sive than one without. Once greenhouse gas emissions are priced by the tax, a plant with CCS may become economical to operate. The appropri-ate rate can be determined by raising the tax until a natural gas plant with CCS or a gasified coal plant with CCS is a more profitable investment than the same plant without CCS.

To avoid Pigouvian taxes, the owners of existing high-carbon fossil fuel plants will extend their plants with CCS technologies. Therefore, in a natural manner, these plant extensions will provide reductions in green-house gas emissions, an avenue for the plant owner to recover more of their original investment, and a soft drawdown of the coal and natural gas electricity generation industries.

CONSERVATION TAX BREAKS AND SUBSIDIES

Although we cannot reduce our electricity use to reach a low-carbon future, we can make the electricity we do use more productive by encour-aging conservation. Governments can promote conservation technologies by creating either tax breaks or subsidies for improving the energy conser-vation of buildings, automobiles, and industries. Installing insulation in buildings is a prime example.

Because conservation can be applied across a broad base of infrastruc-ture, conservation tax breaks and subsidies should also be broad-based. These incentives should be created for both new construction and old buildings. They should be created for a broad section of stakeholders—owners, tenants, and lessors. And they should be created for broad sections of our economy—residential, commercial, and industrial.

APPROPRIATE GOALS

Governments need to match measurable goals with actual problems. Unfortunately, society has become distracted with irrelevant measures. The problem we have is climate change. This phenomenon is caused by greenhouse gas emissions. To stop and reverse the phenomenon, we need to reduce greenhouse gas emissions. Therefore, the appropriate measure is greenhouse gas intensity.

The German government offers a cautionary tale by their establishment of renewables penetration as the relevant measure. Germany is a large-scale renewables role model—in fact, an overachiever in this respect, with an extremely impressive 37 percent renewables penetration. Twenty-five percent are intermittent renewables. Another 13 percent are dispatchable renewables, such as hydro, geothermal, and biomass. After decades of an extremely expensive *Energiewende*, German policy has achieved their aims of a high renewables penetration.

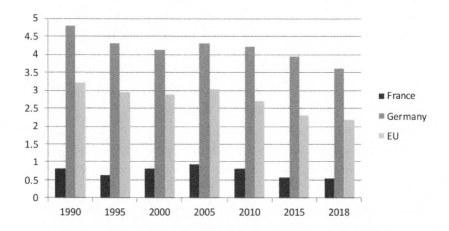

FIGURE 37. Tons greenhouse gas emissions from electricity generation per capita.

However, German greenhouse gas emissions are still on the wrong side of the European Union average. Under the appropriate measure of greenhouse gas intensity, German policy is a failure. A large portion, 37 percent, of German electricity generation remains as coal. Germany externalizes its domestic grid problems to its international neighbors.[2] And their electricity generation greenhouse gas emissions are declining very slowly: a mere 83,000,000 tons reduction over twenty-eight years, from 381,000,000 tons in 1990 to 298,000,000 tons of CO_2 in 2018. Using the Moorburg coal plant emissions of 8,500,000 tons of greenhouse gas each year, it can be calculated that Germany closes one-third of a coal plant each year—only thirty-five to go. Germany must speed up to closing two coal plants a year to meet their promise to stop their ongoing coal catastrophe by 2038. Germany is a laggardly follower in reducing greenhouse gas emissions from electricity generation, despite leading in renewables installation.

	Germany	EU average	France
Greenhouse gas emissions	3.60 tCO$_2$/capita	2.19 tCO$_2$/capita	0.69 tCO$_2$/capita
Multiples of French per capita	5.39	3.17	1.00
Absolute emissions	298,000,000 tCO$_2$	–	46,000,000 tCO$_2$

TABLE 33. Tons of greenhouse gas emissions from electricity generation per capita.

France has a similar economy, similar weather, and a population only 20 percent smaller than Germany's. French electricity generation has a much lower greenhouse gas intensity than the EU average in both per capita and absolute terms.[3]

German emissions are worse than the European Union average by over 40 percent, five times worse than French emissions per capita, and six times worse in absolute terms. The dismal achievement of German policy is because they use an incorrect measure: the number of renewables installed rather than reductions in greenhouse gas.

COLLECTIVE VERSUS MASS ACTION

Collective action is characterized by a bunch of people organizing society. For example, it might involve electing politicians who are motivated to address energy issues, who will be guided by science in defining appropriate energy policies, and who can pass laws to enact such policies. Society can organize solutions and mobilize efforts such as Pigouvian carbon taxes, fair pricing for microgenerators, and appropriate investment in utility-scale generation.

Mass action, on the other hand, is characterized by a bunch of people doing the same thing. For example, if a million people installed solar PV on their rooftops—which seems like a lot—then 0.013 percent of the global population has slightly cleaner electricity for their homes. This is not much progress and may even be retrograde, because society will have to deal with new issues, such as peak solar at noon creating problems at the distribution level of the grid, households without solar PV subsidizing households with solar PV, and underinvestment in utility-scale generation supporting the economy and jobs. Mass action is the aggregate result of an unorganized mob.

Collective action is the result of society's political and economic mobilization, and it is our only way forward that results in a realistic and effective low-carbon energy future.

CHAPTER 15

APPLYING THE MODEL

OF COURSE, YOU MAY disagree in whole or in part with my proposed solution here, but as David Mackay[1] points out, any solution must be pro arithmetic and must add up! I hope that as you seek your own solution, the concepts and valuation model explored in this book will enable you to realize which characteristics a practical solution contains.

SET THE GOAL TO REDUCE GREENHOUSE GAS EMISSIONS

Society must set the goal to reduce carbon dioxide emissions. We must stop irrelevant measures, such as measuring the penetration of renewables installed. Reducing greenhouse gas intensity is the appropriate and relevant measure. How many grams of carbon dioxide are produced per kilowatt-hour (gCO_2/kWh)? Calculating that figure and then lowering it will lead to real positive change.

This goal must be seen within the context of expanding electricity generation to support all of humanity at first-world power levels. Whether you view yourself as a ratepayer or a taxpayer, a politician or a voter, a businessman or a laborer, we all live on the same planet. We will all be much happier in a low-carbon, highly electrified future.

CARBON TAXES

Governments should impose a carbon dioxide emission tax, with step-function increments, providing a long-term, stable, and predictable price signal. Such a tax is the simplest and most neutral solution of ensuring that penalties and subsidies are applied in productive directions.

SPECIAL COAL TAXES

The immense damage coal wreaks, because it includes emissions of particulates, sulfur oxides, nitrogen oxides, radioactive particulates, mining wastes, and ash tailings piles deserves an extra consumption tax over and above carbon taxes.

SPECIAL NATURAL GAS TAXES

The use of natural gas to generate electricity must be discouraged. The need for natural gas in industry, transportation, and industrial heat processes will take time to replace. However, natural gas used in electricity generation deserves an extra consumption tax over and above carbon taxes.

INCENTIVES FOR ELECTRICITY CONSERVATION

Conservation is a huge component of realizing a low-carbon future. This is true irrespective of which technologies are used. Conserved electricity can be channeled into other productive uses, such as retiring coal and natural gas plants faster.

Incentives may be structured as tax breaks or as direct subsidies. Furthermore, there are indirect subsidies, such as government-sponsored research into more efficient lighting, appliances, tools, insulation, and other materials.

ENCOURAGING ENERGY-CONSERVING BUILDINGS

Buildings can be improved by adding incentives for owning buildings with low thermal signatures and low use of utility power. This will encourage installation of insulation, heat pumps, micro concentrated solar, and micro solar PV.

The incentive should be oriented toward ownership rather than only new construction. This encourages investment in old buildings as well as new. It also encourages lessors to invest in buildings that a lessee utilizes. A secondary incentive should be provided for tenants and lessees who seek out high energy-conserving buildings to encourage lessors' investment.

COMMERCIAL MICRO SOLAR PV

Micro solar PV will find its niche within cities where commercial rooftops and parking lots go otherwise unused. This is particularly relevant as the pairing of commercial spaces and solar PV tends to match the daytime use of commercial properties with their daytime energy usage pattern.

By matching the generation and usage pattern of solar PV, restricting smart grid management to building-level concerns, and avoiding over-building utility-scale plants, solar PV can serve an important and useful function in a sustainable low-carbon future without too much impact from the negatives of its high intermittency and low capacity factor.

RESIDENTIAL MICRO CONCENTRATED SOLAR

Micro concentrated solar will find its niche within cities where residential rooftops go otherwise unused. This is particularly relevant as the pairing of residential spaces and concentrated solar heating tends to match the daily storage of heat with evening cooking, bathing, washing, and heating needs. Integrating concentrated solar with heat pumps provides economies of scope in the equipment installation.

By matching the generation of concentrated solar and the consumption pattern of heat and by avoiding the lack of smart grid installation by using heat directly, concentrated solar can serve an important and useful function in a sustainable low-carbon future without the negative impact of its high intermittency and low capacity factor.

UTILITY-SCALE POWER

Existing hydropower, geothermal, and Gen II/III nuclear plants should be continually upgraded and improved. New riverine hydropower should be pursued at the country and regional level, although not at the cost of an ecological disaster. Although run-of-river turbines do not contribute much at scale, they are relatively easy to permit and install. Geothermal should be pursued at the regional level.

GEN IV NUCLEAR AS THE PRIMARY POWER SOURCE

The majority of utility-scale power is to be provided by a titanic scale build-out of Gen IV nuclear plants. The high capacity factors, zero intermittency, and small geographic footprint permit great flexibility in siting and usage. Future siting considerations will take into account process heat customers from hydrogen generators, cement makers, ammonia makers, biofuel generators, distillation plants, waste incinerators, and other industries. Leftover low-grade waste heat can be used by water and sewage treatment plants.

WIND ENERGY AS A SECONDARY RESOURCE

Wind can be used at the utility scale to supplement time-independent energy loads. Wind energy will be most useful to the sectors in which it is easy to adjust the demand load, such as agriculture—serendipitously where the electricity is generated. Wind energy may be coupled with load-following Gen IV nuclear plants' process heat to produce carbon-neutral chemical fuels for transport.

By using wind energy for time-independent energy loads, the negative impact of its high intermittency and low capacity factor can be overcome. The more managed demand load exists in a smart grid, such as directing wind to time-independent tasks, the more wind energy we can profitably use.

UPGRADE THE GRID

Investments in a smart grid will pay future dividends whatever the mix of power sources and wherever the locations of the generating plants and demand loads are. A fully smart grid will enable society to be more efficient and achieve decarbonization faster. In particular, demand shifting will allow more targeted investments in physical grid components such as wires, transformers, and switching stations.

STORAGE RESEARCH AND ADDITIONS

Investments in storage research will pay future dividends, whatever the mix of power sources and wherever the locations of the generating plants and demand loads are. Additional storage capacity should be pursued—but at a moderate pace. The current leading storage mechanism, pumped hydro, can have devastating ecological impacts if not thoroughly studied and carefully built. Compressed air storage can have impacts similar to those of fracking on geologic formations, and it is clear that we do not fully understand the full impact of fracking technology. Liquid air storage is very promising due to its expansion potential, but it is currently deployed very narrowly. Without some major breakthrough, utility-scale chemical battery storage is likely to remain a niche, fulfilling smart grid goals rather than attaining the long-term and seasonal storage needed to back intermittent energy resources.

Other storage solutions have less general applicability than chemical batteries. Among these are mechanical flywheels, gravitic sledges, and electric supercapacitors. These solutions will remain in niche applications, likely restricted to specific locations and purposes, and are extremely unlikely to become mainstream installations. Thermal storage using molten salts has a role, but it will likely be restricted to enable load-following power plants rather than as a separate utility-scale storage solution.

BEYOND ELECTRICITY: DECARBONIZING INDUSTRY AND TRANSPORT

In order to achieve a low-carbon society, we need to also address the other energy sectors[2] contributing to greenhouse gas emissions—mainly industry and transport. Industry is not very amenable to electrification as it mostly uses energy as raw heat. Transportation is somewhat amenable to electrification. However, the more exciting opportunity is replacing high-carbon fossil fuels with low-carbon synthetic fuels, allowing us to continue using existing combustion-based assets.

INDUSTRIAL PROCESS HEAT

Industries that switch to low-carbon Gen IV nuclear process heat sources should be rewarded with investment tax breaks. Particularly energy intensive industries, such as cement and ammonia production, should be prodded with the stick of an extra high-carbon-intensity tax, over and above its taxation on carbon dioxide emissions. These industries would then have both carrots and sticks driving them toward the new investments required.

HYDROGEN GENERATION

High process heat enables the generation of hydrogen from water more cheaply than simple electrolysis. This works in three pathways to provide cheap and high-efficiency hydrogen: Simply by using heat directly, less energy is wasted in the inefficiencies of using heat to generate electricity and then in the inefficiencies of using electricity to generate process heat. Raising the temperature of water improves the efficiency of electrolysis itself. It allows us to avoid the current high-carbon method of generating hydrogen, which is to partially burn methane to produce hydrogen gas, plus carbon monoxide or carbon dioxide.

A plentiful supply of hydrogen enables the biofuels and carbon-neutral fuels described next. In short, we need lots of hydrogen to make low-carbon transportation fuels, and Gen IV nuclear gives us a three-for-one return on avoiding carbon-dioxide emissions to do so.

TRANSPORTATION AND NUCLEAR CARBON-NEUTRAL FUELS

There are several chemical processes that utilize high-temperature process heat and hydrogen to produce carbon-neutral fuels for transport. Gen IV nuclear can provide the high-temperature process heat needed to convert atmospheric carbon dioxide to methane, alkanes, alcohols, and diesel fuels for sustainable transport. Ammonia is another potential carbon-neutral transport fuel that, with sufficient process heat, can be synthesized directly from the air.

TRANSPORTATION AND BIOFUELS

The carbon dioxide and particulate footprint of biofuels can be improved. Through gasification or hydrogenation using the high temperatures of Gen IV nuclear process heat, biomass can be made into alkanes and diesel fuels. Biomass is more useful when packaged for sustainable transport than burnt for electricity generation.

Given these high temperatures, even the poor-quality fuel of garbage used in waste incinerators is a viable feedstock for transport-fuel generation. In addition, gasification or hydrogenation changes the waste profile of the ash tailings from burning garbage. The ash tailings will be smaller in volume and more amenable to further recycling.

DIRECT NUCLEAR MARINE PROPULSION

The American, British, Chinese, French, and Russian navies have demonstrated the viability of nuclear marine propulsion for decades. Large cargo ships use extremely dirty bunker oil, which you may think of as a sludge of liquefied coal. It is so dirty that it has to be melted, literally heated above 100°C until it liquifies, to be piped into the burners. Replacing these highly polluting point sources with simple, sustainable, ship-size Gen IV nuclear power turbines across all cargo ships would reduce global carbon dioxide emissions by 2 to 3 percent in and of itself!

CHAPTER 16

CONCLUSION

OUR GLOBAL SOCIETY requires a titanic amount of electricity to maintain itself. We require an immense amount of electric power to maintain the standard of living in the first world. We require more electricity and process heat to decarbonize transportation and industry. We require even more electric power in order to improve the lives of those living in the third world.

Our society requires low-carbon electricity to avoid the terrible consequences of climate change. As such, we must measure progress on climate change in terms of the amount of emissions reduced.

The policy solutions of Pigouvian taxes on emissions, grid improvements, and energy conservation must be pursued vigorously.

Nuclear power is the primary technology to achieve abundant, low-carbon electricity at titanic scale. Although costly to build, nuclear plants have a high valuation due to the advantages of the dispatchable, high capacity factor, installable, dense, low-carbon power generated. Nuclear plants have been proven to kill coal plants and to reduce fossil fuel dependence quickly and efficiently. Nuclear process heat can decarbonize transportation and industry. The issues of cost and waste can be overcome with Gen IV reactors. The problem of fear must be overcome by our schools and our politicians.

The renewables of wind, solar PV, and concentrated solar are secondary technologies to supplement nuclear power. Wind and solar PV technologies have titanic scale energy, but they do not have titanic scale *power* without Atlas-size storage. Concentrated solar technology has titanic scale power over the hourly time frame, but for longer time frames, it also requires Atlas-size storage. Research and innovation will never overcome the problems of high intermittency, low capacity factors, siting difficulties, and diffuse energy.

Improvements in storage, the addition of storage capacity, and upgrading to a smart grid will improve both nuclear technology and the renewable technologies of wind, solar PV, and concentrated solar. Storage improvements and capacity expansion will be modest. Long-term and seasonal storage will remain below titanic scale. Storage installations will support and improve the grid's operating parameters. Storage will make a wider array of residential storage and electric vehicles possible, but it will not reach the scale of fundamentally altering the generation capacity needs of utilities serving residential, commercial, and industrial consumers within cities. Society would be taking a grand gamble to rely on fundamental advances in storage technology to mitigate the drawbacks of wind, solar PV, and concentrated solar energy.

Hydropower and geothermal power are tertiary technologies due to the siting and installation difficulties of expansion. These technologies will always remain restricted due to geographies and ecological concerns.

MY ANSWER TO RYAN BOYLE

SOLAR PANELS MAY reduce your electricity bill, but solar panels on your roof will not do much to stop climate change. Only 10 to 20 percent of the energy you use is consumed by your home, and most of that is consumed as heat, not electricity. Given that we live in the moderately high latitude of only somewhat sunny Chicago, my suggestions for improving your residential carbon footprint are to do the following:

· Examine the value your house would receive from upgrading and adding insulation.

· Install a thermostat system that can be programmed for automatic adjustments.

· Examine the value your house would get from a heat pump.

· Look into concentrated solar's heat rather than solar PV's electricity.

· Analyze your appliances for replacement, including nonelectric devices, starting with the oldest.

· Vote for and support politicians who both understand the impact of climate change and are for policies that are pro arithmetic, realistic, and of titanic scale.

GLOSSARY

NAMEPLATE CAPACITY

Nameplate capacity is the design generation capacity of the plant. The design takes into account real-world energy losses and frictions. For example, a solar PV farm may be designed to generate 1,000 MW of peak power, given assumptions about its area, composition, latitude, and prevalent weather conditions.

CAPACITY

Capacity is the practical generation capacity of the plant. Capacity will always be lower than nameplate capacity because of plant maintenance, grid maintenance, and grid requests. The variability that intermittent energy sources experience reduces capacity.

CAPACITY FACTOR

The capacity factor is the quotient of capacity and nameplate capacity. The capacity factor may be thought of as the operating uptime of a plant: The higher the capacity factor, the more time the plant spends generating electricity.

It is extremely difficult to raise the capacity factor. We can reduce our maintenance cycle through better manufacturing techniques and operating experience, raising the capacity factor a small amount for any type of plant. This improvement has no effect on intermittency; reducing maintenance to a small amount today will not improve tomorrow's weather. Likewise, an improvement in capacity factor has no effect on efficiency or intermittency.

DISPATCHABILITY

Dispatchability is the ability of a power plant to respond to the demand load based on human direction. Because dispatchable plants conform to human desires, their power output is more valuable.

It is not possible to design a dispatchable power plant using intermittent energy resources, although it is possible to design dispatchable storage charged by intermittent energy.

INTERMITTENCY

Intermittency is the variability of weather-based energy resources. From the capacity factor, we know that in aggregate, our solar PV farm will generate electricity 24.5 percent of the time. It may be sunny, cloudy, or nighttime over the solar farm at this exact moment, so generation at the moment may be 50 percent or 12 percent of nameplate capacity. A wind farm may experience storms, windiness, or calm.

It is impossible to reduce intermittency. We may be able to mitigate intermittency through titanic scale storage or gigantic efforts to shape the demand load through a smart grid. We may also be able to mitigate intermittency through weather forecasting, so that we can request a dispatchable plant to make up the difference.

Intermittency, a short-term random variability of the energy resource (wind or sun) on the order of seconds, minutes, hours has no effect on the capacity factor, a long-term aggregate of the energy farm's (wind turbine or solar panel) generation on the order of months or a year. Likewise, the intermittency of the energy resource does not change the efficiency of a wind turbine or a solar panel design.

EFFICIENCY

Efficiency is the quotient of our designed plant's capacity and a theoretical plant's capacity. For example, a theoretical solar farm will extract 100 percent of the energy of the photons hitting the solar panel. A real-world solar panel is optimized to capture photons of a particular wavelength but fails to capture all photons, the wires experience resistance, and the panel heats up rather than making electricity; all of these physical effects lower efficiency.

It is certainly possible to improve efficiency. We can design better solar panels able to absorb broader ranges of wavelengths and require less maintenance. We can build better solar thermal capture geometries and use better coolants. This efficiency increase reduces the size of the installation and lowers the cost of installation, but it doesn't change the capacity factor or the intermittency. The Sun only shines during the daytime, we cannot control the movement of clouds, and prevalent local weather will remain prevalent.

We can design better wind turbines, larger wind turbines, turbines that require less maintenance. This efficiency increase reduces the size of the installation and lowers the cost of installation, but it doesn't change the capacity factor or the intermittency. We cannot force the wind to blow at the proper speeds when and where we want it to. Efficiency improvements in wind turbines do not result in steady and stable wind at the optimum speed of the turbine.

We can design slightly better turbines for hydropower plants, and we can slightly improve the insulation and turbine equipment for thermal plants (geothermal, nuclear, concentrated solar). However, progress in these areas will be incrementally small because of the maturity of the science and engineering that we already have.

PREDICTION

Prediction of intermittent energy sources mitigates intermittency by providing data on which to make the decisions of whether to use a smart grid to manage demand load (assuming you have a smart grid) and whether to request a dispatchable power plant (by definition, an intermittent plant cannot be dispatched) to reduce or increase generation to meet demand load. Prediction does not change the capacity factor, intermittency, or efficiency of an energy resource.

POWERING THE FLATLANDS
OF BELGIUM

TABLE 34 COLLECTS THE single-provider information for a quick comparison between the basic six low-carbon technologies that might be used to provide the 88,600,000 MWh of electricity and heat Belgium uses each year. Of course, a real-world solution will use a combination of these technologies. Still, the energy-area trade-off shows the obvious choices that a country must make without America's vast areas or Norway's mountains, without Icelandic volcanism, and without similar solar radiance as in California.

Plant	Per-plant annual energy	Number of plants	Percentage of the land area of Belgium	
Boundary Electric Plant	3,587,000 MWh	25	5,300%	dual use
Hellisheidi Geothermal Power Station	2,828,000 MWh	31	–	dual use
Centrale Nucléaire de Cattenom	36,739,000 MWh	2.4	0.03%	single use
Roscoe Wind Farm	2,174,000 MWh	41	53.4%	dual use
Topaz Solar Farm	1,262,000 MWh	69	4.3%	single use, California weather
Solana Generating Station	792,000 MWh	112	2.9%	single use, California weather

TABLE 34. Consolidated comparison of the low-carbon technologies for Belgium.

Hydropower and geothermal plants are simply not possible to site given Belgium's flat and stable geography.

SYNTHETIC FUELS

GASIFICATION AND METHANATION

Gasification converts solid carbon into gaseous hydrogen and carbon dioxide by combining a carbon source (garbage, biomass, wood, coal) with steam at a high temperature. Gasification is better than combustion, because the gases produced are much cleaner burning. The gases produced may be used in fuel cells, and cleaner-burning gas is more amenable to carbon capture and storage. Collecting the residual ash is also much easier than in a standard incinerator. Without gasification, burning garbage and biomass is often worse than burning coal in emissions of particulates and greenhouse gases.

The basic reaction occurs at high temperatures (700°C or higher, where nuclear process heat operates):

$$C + H_2O \rightarrow CO + H_2$$

At this point, we have gasified solid carbon into carbon monoxide and hydrogen. These can be further burned in oxygen or used in fuel cells. Or we can keep the temperature up and add more steam to the carbon monoxide, resulting in methane and carbon dioxide:

$$4CO + 2H_2O \rightarrow 3CO_2 + CH_4$$

Methanation is a similar reaction that directly reduces carbon dioxide to methane by adding hydrogen. This leads to very pure methane without the impurities of natural gas and will be free from the particulates and sulfur oxides of standard combustion. The carbon dioxide may be captured directly from the atmosphere, from the CO_2 dissolved in water, and of course, from the carbon dioxide waste from gasification:

$$3CO_2 + 4H_2 \rightarrow CH_4 + 2H_2O$$

At first glance, these equations may look like a chemical perpetual-motion machine, but in each case, both process heat and hydrogen, gaseous or from water, are continually being added.

CARBON-NEUTRAL FUELS

Atmospherically derived carbon-based fuels are interesting because we can convert carbon dioxide dissolved in the world's oceans into alkanes, alcohols, and diesel fuels through chemical reactions at high temperatures.

Methane can be produced via methanation as described previously. Methane can be used directly or can be used to produce longer-chain alkanes such as propane and butane and alcohols such as methanol and butanol. Long-chain alkanes and alcohols can be stored at lower pressures than methane, improving both safety and storage density.

Diesels are desirable because they may act as jet fuels for aviation transport. Similar to methanation, diesels may be synthesized by the addition of process heat and hydrogen. The resulting products will include dimethyl ether or other diesel substitutes.

HYDROGEN

Hydrogen may be generated by splitting water into hydrogen and oxygen. There are four basic methods of splitting water:

1. Electrolysis (the least energy-efficient method)
2. High-temperature heat
3. High-temperature electrolysis
4. High-temperature chemical catalysis cycles, such as the sulfur–iodine cycle

Hydrogen is a poor primary fuel, because it is difficult to store and transport. However, it can be used as a feedstock for ammonia or the carbon-based fuels described previously, fuels that are simple to store and transport because of their high density at atmospheric pressure.

ZERO-CARBON AMMONIA

Ammonia (NH_3) is a great synthetic fuel to pursue, because we already have a great deal of knowledge about how to produce it using materials from the atmosphere—nitrogen and hydrogen. The chemistry and engineering of the Haber process is well researched and understood. A large number of ammonia plants for agricultural fertilizer have already been built across the globe.

Ammonia is a great way to store hydrogen. It liquefies at room temperature and only a few atmospheres pressure, and it may be stored dissolved in water. Therefore, ammonia can be stored at lower pressures than hydrogen, improving both safety and storage density.

In addition to ammonia's ability to store hydrogen for fuel cells, it can be used in fuel cells directly or burned in a modified internal combustion engine. Of course, because ammonia contains no carbon, its combustion products contain no carbon dioxide. It is a zero-carbon fuel!

STORAGE

OPTIMISM IS NOT A STRATEGY

Daniel Kammen, a noted scientist with many positive contributions to society, is, unfortunately, ridiculously optimistic. His optimism, and the optimism of those who share his views, is dangerously misleading.

Optimistic statements about future scientific and engineering progress are not science! Precognition, divination, and fortune-telling are the realm of the mystic. Grant applications for blue-sky science may be an appropriate arena for predicting the future. Investments in practical low-carbon generation plants and equipment are not. When saving our planet from climate change, society's investment in plants and equipment must be in known solutions.

The following principles of caution apply: We should wager small amounts of money on fundamental and potentially disruptive research. We should invest larger amounts of money on existing science and engineering with expansion potential. We should invest the largest amounts of money in building existing low-carbon power plant designs.

To be clear, we will improve storage technologies. We will make scientific and engineering progress in storage. What we are unlikely to do is to make fundamental breakthroughs in these technologies, because such breakthroughs would indicate that our fundamental understanding of chemistry and physics has changed.

What we are completely unable to do is to make scientific progress according to a timetable. America's pledge to place a man on the moon in ten years was an example of an engineering and financial timetable—all the technologies existed at the start if we were willing to pay the costs. The scientific innovation that occurred during those ten years made the effort simpler, cheaper, and easier, but it could have been done with a more complicated, expensive, and difficult solution. The timetable of future scientific progress is unknowable.

GRAVITIC STORAGE

Gravity-based storage consists of pumped hydro, weights traveling up and down in mineshafts, railcars on mountains, and the like. Gravitic storage has almost no disruptive potential, and it has extremely little expansion potential.

There will be incremental progress in developing more efficient water pumps, lower-friction pulleys, and cables for weights in mineshafts or on rails. There will be progress in proper siting and building cost reductions. However, there are simply not going to be disruptive breakthroughs that reduce recharge time or raise round-trip charge efficiency. Understand that there is no fundamentally new science here:

- Water has a known mass and a known viscosity.
- Weights have a known mass, and steel cable has a known tension and friction coefficient.
- The Earth has a known and constant gravitational field.
- The equations of gravitational potential energy for storage and kinetic energy for discharge are well understood.

Currently, pumped hydro is oriented toward demand peak shaving and grid stability. Pumped hydro storage is the most common type, contributing about 97 percent of all storage. Similar to hydropower plants, there are a limited number of sites on the planet that have a very small watershed, that have as great a difference between the top and bottom as possible, and that have appropriate places for two reservoirs. The good news is that, although the number of useful sites is limited, humanity has yet to find and exploit all of those sites.

Practically speaking, gravitic storage is not going to meet the titanic scale needed for daily or weekly storage, let alone seasonal storage. There are a finite number of suitable high-low reservoir sites for pumped hydro. There are a finite number of suitable mineshafts in which to install weights. Sending railcars up mountains is a nifty idea for a particular location in the mountains of Norway, but it simply isn't scalable to the plains of Belgium any more than pumped hydro is. Contemplate the implications of the fact that the highest point in Belgium is the Signal de Botrange, at 694 m that, with the aid of a wooden platform, has been raised to 700 m (2,296 feet).

MECHANICAL STORAGE

Mechanical storage consists of compressed air, flywheels, and the like. Mechanical storage has almost no disruptive potential and extremely little expansion potential.

Similarly, there will be incremental progress in developing more efficient air turbines and space-age lubricants and bearings for flywheels. Again, there will be progress in proper siting and building cost reductions. However, there are simply not going to be fundamental breakthroughs that reduce recharge time or raise round-trip charge efficiency. Understand that there is no fundamentally new science here: Compressed air needs to be pumped into unused mineshafts (those not used by gravitic storage), fracked earth boreholes (like natural gas fracking), and giant underwater balloons. Liquefied air plants can achieve lower volumes than compressed air using more energy. Flywheels can only be so large in mass and volume due to the physical stresses incurred. And the equations of pressurized air or rotational potential energy for storage and kinetic energy for discharge are well understood.

Practically speaking, mechanical storage is not going to meet the titanic scale needed for weekly storage, let alone seasonal storage. Plant siting is less of a problem than gravitic storage, but plant sizing is contrariwise even more difficult. Compressed air plants require huge volumes, and flywheels require giant masses at the scale needed. Liquefied air plants have lower volumes than compressed air plants, but volume reduction does not change the laws of physics.

THERMAL STORAGE

Thermal storage consists of heating coolants, such as molten salt, and then retrieving energy through a heat engine, either a turbine or a Stirling engine. Thermal storage has almost no disruptive potential and very modest expansion potential.

Short-term storage for nuclear or concentrated solar generation suffers little performance degradation, because they are already creating heat as the primary product. Thermal storage is terrible for hydropower, wind, or solar PV generation because efficiency suffers an immediate hit in the conversion of high-grade electric power to low-grade heat energy.

Thermal storage is only useful for short-term storage because, with our nuclear or concentrated solar plants running full out, we can only absorb as much heat as the plant's capacity factor permits. After that, if we add more coolant mass, we need to add more nuclear or concentrated solar capacity. There is no free lunch. There is no fundamentally new science here: Each coolant (water, organics, salts, metals) has known heat capacities. The heat stored is mass dependent; more coolant means more stored heat.[1]

Practically, thermal storage is not going to meet the titanic scale needed for weekly storage, let alone seasonal storage. Storage plant siting is pretty much confined to the plant boundaries itself, because of heat losses from long pipelines, the large mass of the coolant that must be pumped, and the need for the heat engine to produce electricity.

CHEMICAL BATTERY STORAGE

Battery-based storage consists of materials with differing electrochemical potentials being separated during the charging phase and brought together during the discharging phase. Examples are lithium-ion, vanadium-ion, brine-based, and liquid metal-antimony batteries. Chemical battery storage has almost no disruptive potential and very modest expansion potential. Globally, there is approximately 7,840 MWh of chemical battery storage, about 1.4 percent of total storage.[2]

The amazing progress in batteries must be tempered with the knowledge that the relative electrochemical potentials of each element and many small molecules are well known. The electricity stored is mass dependent; more stuff means more juice.

Taking the lightest possible battery, constructed of elemental lithium, we quickly see that there is a maximum limit to the power density of batteries.[3] Batteries at that power density limit can only store more energy by increasing their mass. Naturally, lithium has a physical density limit, and therefore, to store the energy desired, we must increase the volume. There is a maximum power density, mass, and volume limit for most applications— air transport immediately springs to mind. It is extremely unlikely that we will ever see a much lower mass, much lower volume, much higher power density battery than the lithium-ion cell. Cell packaging, cell membranes, solid-state cells, cell management, and cell electrolytes will all experience incremental improvement. We will see progress, but we will probably not see fundamental, paradigm-shifting breakthroughs.

CHEMICAL FUEL STORAGE

Fuel-based storage consists of applying energy to create liquid fuels that can be oxidized by burning in engines or direct oxidation in fuel cells.

Chemical fuel storage has modest disruptive potential and good expansion potential.

Chemical fuels can be used to store energy during both short and long periods of time. Although chemical fuels may be used for electricity generation, these fuels are better used for the purpose of decarbonizing transportation.

Applying energy from nuclear, wind, solar PV, or concentrated solar plants, we can produce hydrogen, zero-carbon ammonia, carbon-neutral methane and higher alkanes, and carbon-neutral diesel fuels such as dimethyl ether.

The major upside of chemical fuel storage is the ability to reuse the existing fossil fuel infrastructure to manage the fuels. Like the other storage options, such as chemical batteries, liquid fuels are mass and volume dependent. However, liquid fuels have energy densities that are over an order of magnitude more favorable—between nine and nineteen times denser—and so the volume issue becomes less relevant.

Fuel	Volumetric energy density (Wh/liter)	Density multiple over lithium-ion batteries
gasoline	9,700	19
methane	6,400	13
ammonia	4,325	9
lithium-ion battery	250–693	1

TABLE 35. Chemical storage densities.

The disruptive potential of chemical fuel storage comes from research into new chemical reactions to produce the fuels. The expansion potential comes from research into the improvement of existing production pathways through new catalysts, reaction steps, and reaction conditions.

SUPERCAPACITOR STORAGE

Capacitor-based storage consists of storing the electric charge itself in materials. Capacitor storage has modest disruptive potential and modest expansion potential.

As with all the mass-dependent storage solutions—gravitic, mechanical, thermal, and chemical batteries—titanic scale capacitors are impractical, because they occupy very large volumes. Supercapacitors dramatically improve the power density, but it is still an order of magnitude lower than chemical fuel storage, and capacitance still has a linear relationship with mass.

However, for their niche applications, supercapacitors have the potential to be disruptive because newer supercapacitor materials are researched and created. The expansion of supercapacitors has modest gains to be made in fabrication and application.

GRID-LEVEL STORAGE PRICES AND THE RETURN ON ASSETS

Grid-level storage over long periods of time is extremely difficult from the technical point of view. However, with super-cheap storage, perhaps we can incur large and fantastic inefficiencies and simply pay our way through the problem. The following scenario will soon make it obvious that current storage would need to be not just ten times cheaper but *hundreds* of times cheaper.

Examining the scenario of storing summer solar for winter demand, we will optimistically set the price of storage to the price of coal and presume both are priced at $0.02/kWh.[4] Wow! This is truly super-cheap storage! Our customer requires one kilowatt every hour. Between June and January, coal sells 4,383 kWh of electricity, or $87.66 worth of juice. Storage saved 1 kWh in June and sold 1 kWh in January for $0.02 in revenue.

We begin to see the poor economics of storage. Coal covers maintenance, operation, fuel, debt service, and profit with revenues of $87.66. Storage must cover maintenance, operation, input energy from the solar farm, debt service, and profit with revenues of $0.02. Six-month storage at a price of $0.02/kWh is, in fact, completely absurd!

Changing the scenario from unrealistic seasonal storage to a shorter-term storage scenario, we find a smaller but similar problem. Over twenty-four hours, coal has revenues of forty-eight cents. Storage will take twelve hours of input electricity from the solar farm to sell twelve hours of storage at night for revenues of twenty-four cents.

Of course, these are revenues, not profits. We realize that the more complicated coal plant would have set the price above $0.02/kWh if it could not make a profit at that price. However, we are not concerned with the profitability or costs of an individual plant, but rather with the aggregate costs to society. A dispatchable coal plant is a single asset to construct, maintain, and operate.

Storage is the third asset to construct, maintain, and operate! The first asset is the solar farm to provide the customer's energy during daylight hours, and because the law of conservation of energy must be obeyed, when its energy is sent to the customer, it cannot have a dual purpose of supplying energy to top up storage. The second asset is the solar farm to generate energy to place into storage to provide the customer's energy during nighttime hours. Using wind rather than solar, or a combination of the two, results in a similar story—because conservation of energy is not a law that can be broken.

Won't storage improve the return on assets by solving curtailment issues? Solar PV and wind could solve curtailment on their own if they built their own storage, converting their intermittent energy farms into dispatchable power plants. In contrast, semi-dispatchable concentrated solar does not suffer from curtailment, because it builds storage that is inherently sized for the hourly timescale. Separate storage uses other people's assets to solve the intermittency problem that solar PV and wind energy farms create. Curtailment is the grid pushing back. However, given that the current existing rules are such that society must solve these problems to the benefit[5] of the ones creating it, curtailed energy farms are simply overbuilt first and second assets; assets which wait on someone to build that third storage asset. In other words, yes, a terrible return on intermittent assets can be brought up to merely a poor return by the addition of the storage component.

Long-term or seasonal storage based on intermittency requires three assets: the first intermittent energy farm, the second intermittent energy farm, and the storage itself. Unless the marginal price is wildly disparate, a dispatchable power plant will always have a better return on assets to society.

Storage has been shown to be a useful and superior asset to a natural gas peaker plant. However, those situations are for storage of only a few hours, with input electricity bought at low cost and output electricity sold at a high price coupled with the capital and operational savings from *not building* a peaker plant. The lesson is that economical storage deals with short-term, predictable, hourly time-cycle demand loads. The lesson is not that storage is practical for long-term, seasonal, or dynamic demand loads.

DISTRIBUTED STORAGE WITH ELECTRIC VEHICLE CHARGING

The International Renewable Energy Agency (IRENA) has recently published a report that describes using the collective storage capacity of 1,000,000,000 electric vehicles to provide large-scale storage of 14,000,000 MWh in 2050.[6]

This idea is fascinating, and it does have merit, but we must temper optimistic enthusiasm with some practical realities. 14,000,000 MWh of storage adds 156 percent more storage to the 9,000,000 MWh of existing storage. This brings storage from 0.038 percent of global electric capacity to 0.097 percent!

To take advantage of this extra storage, a billion electric vehicles must be built over the next twenty years, which is doable, but those electric vehicles will face competition from internal-combustion vehicles. We also have to build close to a billion or more two-way charging stations at homes and offices. We must then coordinate these two billion assets to charge and dispatch storage. As a closer reading of the report indicates, this will shave the peaks and fill in the valleys of electricity generation. It will not provide long-term or seasonal storage capacity.

The IRENA report vividly illustrates that the problems of storage have few solutions to be optimistic about.

WIND AND SOLAR ENERGY

Empirical evidence shows us that weather is very dynamic. Sometimes, for days or weeks, the wind doesn't blow hard enough to even turn wind turbines. Storms can blow too hard, shutting wind turbines down. Clouds can hide the Sun, preventing solar PV or concentrated solar for days and weeks at a time during winter. The weather can enter *dunkelflaute* periods, when neither wind nor the Sun is providing energy. Eventually, the difference in energy is made up in the aggregate to the constraint of the capacity factor, but without storage, this cannot be converted to a power supply.

Without storage, we need quadruple, quintuple, and even sextuple anticorrelated facilities hundreds of kilometers apart, with titanic-scale transmission flow along the grid. With storage, scaled proportionate to titanic levels storing days, weeks, even months of energy, we merely need three, four, or five times the number of facilities hundreds of kilometers apart, with titanic scale transmission flows.

Extra plants, extra storage, and extra transmission result in a low return on assets. When intermittency is taken into account, then these energy farms sometimes generate electricity simultaneously, and sometimes they are idle simultaneously; the entire system has intermittency that must also be prepared for. The additional issue is what level of intermittency is acceptable. How many seconds or minutes of insufficient power are acceptable?

HURRICANE SCENARIO COMPARING DISPATCHABLE POWER TO INTERMITTENT ENERGY PLANTS

Our ideal scenario cities are geographically distributed across the country, share energy instantly without transmission line losses, and have a demand load of 900 MW each.[7] The intermittent energy farms are a blend of wind and solar with a blended capacity factor of 30 percent. This scenario assumes the most favorable intermittency, with its impact to generation zero, the plants evenly producing at their average capacity factor.

When a hurricane hits Miami, for a week the winds are too high for wind turbines, and the clouds cover solar panels. Therefore, Miami needs to import 900 MW or suffer a blackout.

9 assets	San Diego	Houston	Miami
	3 blended farms	3 blended farms	3 blended farms
Nameplate capacity	3,000 MW	3,000 MW	3,000 MW
Average real capacity	900 MW	900 MW	900 MW
Intermittency losses	pretend zero	pretend zero	pretend zero
Normal curtailment	0 MW	0 MW	0 MW
Storage	–	–	–
Required grid improvements	cost of smart grid	cost of smart grid	cost of smart grid
Result	900 MW OK!	900 MW OK!	blackout during hurricane

TABLE 36. Pretending zero-intermittency energy, no extra capacity, Miami goes dark during hurricane.

One solution is to invest in extra generation capacity for our three cities. Each city needs 1.5 energy farms to support the other two cities when one city's plants are offline, for a total of 13.5 blended farms. 1,350 MW total is idle or curtailed unless there is an emergency and so represents a loss of economic value. We normally curtail 450 MW per city, but when Miami requires power due to a hurricane, we export 450 MW each from San Diego and Houston and import 900 MW into Miami, thus keeping the city lit as shown in Table 37.

	San Diego	Houston	Miami
13.5 assets	4.5 blended farms	4.5 blended farms	4.5 blended farms
Nameplate capacity	4,500 MW	4,500 MW	4,500 MW
Average real capacity	1,350 MW	1,350 MW	1,350 MW
Intermittency losses	pretend zero	pretend zero	pretend zero
Normal curtailment	-450 MW	-450 MW	-450 MW
Transfers during hurricane	-450 MW	-450 MW	+900 MW
Storage	-	-	-
Required grid improvements	cost of smart grid	cost of smart grid	cost of smart grid
Result	900 MW OK!	900 MW OK!	900 MW OK!

TABLE 37. Pretending zero-intermittency energy, excess capacity, Miami lit during hurricane.

A second solution is to invest in storage for our three cities. Storage requires extra generation to charge the storage in the first place. To determine how much storage, we must determine how many days of electricity we need, and to determine how much extra generation, we must determine how quickly we charge the storage. We shall estimate that we need seven days of emergency power, for a total of 50,400 MWh storage capacity. We shall recharge the batteries within seven days, for a fourteen-day window between two significant generation events, requiring an extra real-world 300 MW capacity in each city as shown in Table 38.

The questions remaining are these:

· Is seven days coverage enough to make it through an emergency?

· What is the likelihood that a second emergency immediately follows the first?

· What is the likelihood that a second city has an emergency simultaneously?

- How many emergencies are predicted per year?
- Is it physically possible to build 50,400 MWh of storage?

50,400 MWh is already twice the capacity of the largest real-world storage installation in the world, the Bath County Pumped Storage Station, with 24,000 MWh of energy capacity. Economically, the extra three energy farms and the three storage plants are mostly idle assets.

None of the scenarios illustrate the intermittency of generation—how much of the supply is either too much or too little—and will it balance out. In the real world, the two assumptions of generating energy evenly at the capacity factor rate and that intermittency is zero are ridiculous. Once intermittency is not assumed to be zero, these scenarios require more assets to provide a fully stable electricity supply. The best that can be done is to present a range of probabilistic estimates rather than guarantees.

	San Diego	Houston	Miami
15 assets	4 blended farms, 1 storage plant	4 blended farms, 1 storage plant	4 blended farms, 1 storage plant
Nameplate capacity	4,000 MW	4,000 MW	4,000 MW
Average real capacity	1,200 MW	1,200 MW	1,200 MW
Intermittency losses	pretend zero	pretend zero	pretend zero
Normal curtailment (when not recharging storage)	−300 MW	−300 MW	−300 MW
Storage	16,800 MWh	16,800 MWh	16,800 MWh
Transfers during hurricane	−300 MWh −16,800 MWh	−300 MWh −16,800 MWh	+600 MW +33,200 MWh
Required grid improvements	cost of smart grid	cost of smart grid	cost of smart grid
Result	900 MW OK!	900 MW OK!	600 MW + 50,400 MWh = 900 MW OK!

TABLE 38. Pretending zero-intermittency energy, storage added, Miami lit during hurricane.

Table 39 illustrates the equivalent dispatchable scenario using a 1,000 MW Gen IV nuclear plant with a 90 percent capacity factor. To deal with maintenance downtime, it carries the overhead of an eleventh backup reactor shared with ten other nuclear plants. Intermittency is, of course, zero. If a hurricane requires the plant to go offline for hours or days as the storm crosses, the shared eleventh reactor is used to dispatch power just as it does during maintenance windows. Each of the three cities shares the eleventh reactor 10 percent of the time, thus an asset of 0.3 plants is listed under the shared plant.

As the backup is also scheduled for maintenance, we can predict that it will be working when needed. Because of this predictability, the total system capacity factor is essentially 100 percent. In addition, the length of a generation disturbance is irrelevant; the shared plant can keep operating for seven hours, seven days, or seven weeks, and we do not need a range of probabilistic estimates to constrain a fully stable electricity supply.

	San Diego	Houston	Miami	Shared plant
3.3 assets	1 nuclear plant	1 nuclear plant	1 nuclear plant	0.3 nuclear plant
Nameplate capacity	1,000 MW	1,000 MW	1,000 MW	1,000 MW
Average real capacity	900 MW	900 MW	900 MW	900 MW
Intermittency losses	actual zero	actual zero	actual zero	actual zero
Normal curtailment	0 MW	0 MW	0 MW	0 MW
Transfers during hurricane	0 MW	0 MW	+900 MW	−900 MW
Storage	0 MWh	0 MWh	0 MWh	–
Required grid improvements	$0	$0	$0	$0
Result	900 MW OK!	900 MW OK!	900 MW OK!	–

TABLE 39. Dispatchable power, all three cities always lit.

Now we can compare each of these scenarios against each other. As we connect an entire nation's generation resources and demand loads, as we include real-world intermittency projections from empirical data, and as we examine real-world transmission losses, local and dispatchable power becomes more and more desirable.

	Intermittent energy, excess capacity	Intermittent energy, storage added	Dispatchable power
Number of assets	13.5	15	3.3
Nameplate capacity	13,500 MW	12,000 MW	3,000 MW
Average real capacity	4,050 MW	3,600 MW	2,700 MW
Intermittency losses	pretend zero	pretend zero	actual zero
Normal curtailment	−1,350 MW	−900 MW	0 MW
Maximum curtailment	−10,850 MW	−9,300 MW	−300 MW
Storage	–	50,400 MWh	0 MWh
Required grid improvements	cost of smart grid	cost of smart grid	$0

TABLE 40. Comparison of intermittent and dispatchable scenarios.

As we can see from Table 40, dispatchable power has less overhead—fewer plants to build and maintain. It provides an appropriate return on assets: There is no overinvestment in either idle or overproducing plants. Even the shared plant is not curtailed because it is usually replacing some plant under scheduled maintenance. There are no indirect costs; storage or grid investments are not required. And intermittency is avoided: There are no estimates of the frequency or duration of weather events to make during design, construction, and operation.

THE VALUE OF PREDICTABILITY

A DISPATCHABLE ENERGY source is a power source. An intermittent energy source is a low-grade energy source.

	Dispatchable	Intermittent
Capacity factor near 100%	high-grade power source has high value	high-grade energy source (none known at scale)
Capacity factor lower than 50%	low-grade power source has moderate value	low-grade energy source has low value

TABLE 41. Value of predictable power.

Even with a high capacity factor, the predictability of power is a huge benefit. The extreme example of providing reliable power to the surgeon in the operating theater is obvious—so much so that hospitals have batteries and backup generators for power outage contingencies.[1]

Let us examine the scenario of a manufacturer. The manufacturer has an industrial contract with the power utility, which is very different from the retail contract because many industrial contracts permit the utility to undersupply power. Therefore, at peak demand load, such as heat waves, cold snaps, the World Cup, the World Series, and the Super Bowl, the manufacturer understands that power is likely to be withdrawn to support the grid. This is an inconvenience, but the manufacturer is still able to plan contingencies and is compensated for the inconvenience.

What the manufacturer has not signed up for is the supply of unpredictable intermittent power. If machines, tools, computers, and sensors capriciously stop when a cloud passes over a solar array, then the manufacturer may not complete their must-ship orders. Late orders mean fees and penalties assessed and less business with that customer in the future.

We can advise the manufacturer to make capital investments in on-site batteries and backup generators, but then that means the utility and grid are providing less value, not more. At a certain point, the manufacturer will just build, own, and operate a private power plant. Specialization is economically more efficient; therefore, our economy has stratified into manufacturers that make widgets and utilities that generate electric power. To expect manufacturers to be power plant owners and operators is the outcome of reliance on intermittent energy resources, which is ridiculous—but all too probable.

High predictability is required even in noncritical situations. My friend Dave is a retiree who is in no hurry to be anywhere or do anything. Dave likes to putter around the house doing odd jobs. However, if the electricity out of the socket is not reliable, then the power drills and saws that Dave uses are not reliable. Although Dave's workshop is not listed as a critical resource by the utility company's emergency planners, it is critical that the power is predictably on or off *at the moment of use*. If the power drill suddenly stops or the power saw suddenly starts up again, Dave may suddenly have to ride a high-speed ambulance to the hospital mentioned previously.

INTERMITTENCY DESTROYS VALUE

Small levels of intermittent wind and solar energy may cancel each other out, providing a stable power supply through anticorrelated assets. However, weather is broadly correlated in space and time. Therefore, high levels of intermittent wind and solar energy may reinforce each other, producing wild swings in over-generation (curtailment) or under-generation (blackouts). This is most easily seen with solar PV: Doubling the number of panels in nearby or longitudinally separated solar farms does not affect generation at sunrise or sunset much, but it will double the generation at noon. Solar generation is correlated with the Sun, not with the demand load.

It is possible to use models of weather to predict the wind and solar energy available to the grid. We need only understand the fundamental truth that predictions of intermittent energy do not change the weather or enable us to produce the correct amount of electricity. Predictions enable us to compensate for the shortfalls of intermittent energy by directing dispatchable power.

Basic economics implies that the intermittent wind and solar energy generation will be assigned lower prices in an attempt to discourage electricity production when they oversupply the grid. Lower prices reduce the incentive for an investor to finance and build more assets, as would be needed in a more reliable but anticorrelated set of renewable generation facilities. Even if the energy farms are usually anticorrelated, they will at times collectively be generating, oversupplying the grid and lowering prices.

Of course, we can pay more to encourage electricity production when intermittent energy undersupplies the grid, but a higher unit price won't magically change the weather.[2] The Earth does not have a bank account to accept the payment! When the undersupply starts and prices rise, the energy farms are not generating the electricity that would capture those higher prices. An investor interested in a method to guarantee electricity production during periods of high unit prices will build a dispatchable power plant. This explains the popularity of pairing high-carbon natural gas power plants with wind or solar energy farms.

However, now that the investor has a larger set of assets, greater capital costs must be repaid over the same demand load. Although a portion of the demand load can be moved about in time, fundamentally demand load does not increase due to generation increases. The cart remains behind the horse. Without bulk energy storage, high penetration of highly intermittent resources affects the economics of electricity supply negatively.

THE LAW OF LARGE NUMBERS

The law of large numbers is commonly misunderstood. This law tells us that a mean calculated from many data points will be more precise than a mean calculated from a few. The law tells us that as the number of data points increase, the mean will not change much. The law does not tell us that the data points will cluster around the mean. The law does not tell us that as the data points change over time, the mean will not change. The law of large numbers holds true. It is the application of the law that is often incorrect.

For example, a coin flip does not cluster around the mean. The *mean outcome* of a coin flip is half heads and half tails. The *actual outcome* of a coin flip is either heads or tails.

Population size	Number of one-year-olds	Number of seventy-nine-year-olds	Mean age (years)
6	5	1	14
200,000	100,000	100,000	40

TABLE 42. The law of large numbers does not indicate the mean has predictive capability.

The mean, or average, is a mathematical construct and does not imply the usual or steady-state condition.

EMPIRICAL CORRELATION OF INTERMITTENCY

ERCOT data of wind generation in Texas allows an empirical test of the law of large numbers. Texas is a fairly windy part of the United States and has a high penetration of wind energy, with thousands of wind turbines spread across many, many wind farms producing a hefty 21,200 MW of electricity. Texas is an extremely large area of 696,241 km² (268,581 square miles), or nation size, being larger than 150 of the roughly 190 countries on the planet.

We do not see wind generation approximately equal to the annual mean across this large area and across time. We see intermittency at the monthly and daily timescales (Figure 18), as we have already seen at the hourly timescale (Figures 20 and 21). A large number of wind turbines across a large area does not reduce intermittency because the weather systems that drive the turbines act as a correlating factor across large areas.

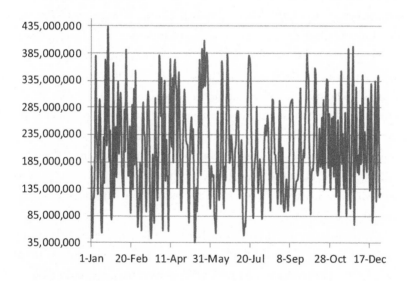

FIGURE 38. Texas wind shows intermittency across the year.

Is the law inapplicable because the mean generation is changing in time over such a large number of wind turbines across such a large area? No. The law tells us that adding or subtracting a wind turbine from the calculation at any instant in time will not change the outcome of that instant's mean generation very much. The mean is calculated from the wind's strength; it is not a setting on the turbines to which the wind responds.

The insight needed is that the grid operator needs to balance the instantaneous demand with *matching* instantaneous electricity generation, not the calculated mean generation. Having a lot of wind turbines guarantees the precision of the calculated mean generation, not that the generation will be close to the demand.

CONCENTRATED SOLAR PROCESS HEAT

Concentrated solar produces heat that can be used for industrial processes, much like nuclear heat. However, unlike the vast potential of nuclear process heat, concentrated solar has several drawbacks. Concentrated solar typically operates in the 250–600°C range, although this can be improved.[3] The Sun only shines for half the day; most industrial processes need constant heat over a twenty-four-hour period. Siting is poor: Prime concentrated solar sites tend not to have nearby clusters of existing industries, and existing industry sites tend to be subprime concentrated solar sites.

Industrial processes that would use process heat, such as cement manufacturing or other chemical production, operate on a 24/7 schedule. The always-on operation is partly to maximize the return on assets of the manufacturing plant, but mostly due to the technical limitations of the process itself. For example, the process chemistry may require sustained temperatures for hours or days.

Let us examine a scenario of a chemical plant in a perfectly sunny geography operating at 1,000°C. The chemical plant requires 1,000 MW_th of heat every hour, or 1,000 MW_t power. In order to do this, we build a 1,000 MW_t power concentrated solar farm.[4] This concentrated solar farm supplies 12,000 MW_th of heat from sunup at 6:00 a.m. to sundown at 6:00 p.m. However, the heat supplied to the chemical plant is only online from 8:00 a.m. to 6:00 p.m., because it takes several hours to heat up the hundreds of tons of coolant, the thousands of meters of piping, and the primary reactor vessels of the chemical plant.

We need to build a second concentrated solar farm that works from sunup to sundown to create a reserve of heat for nighttime use. We cannot use the first concentrated solar farm to heat the extra coolant, and we are not able to produce energy out of thin air! To collect more energy, we must have more solar farm. With the extra heat, the coolant, piping, and reactor vessels can be kept at operating temperatures.

Of course, no geography has perfectly sunny days. To accommodate half-sunny days, we therefore need four concentrated solar farms in total—two, each operating at half capacity, to supply today's thermal power and two, each operating at half capacity, to supply tonight's thermal power. Should we accommodate quarter-sunny days or eighth-sunny days? We can cut to the quick: How much capacity do we need to supply 1,000 MW_t on average demand, at minimum demand, and how long does storage have to last?

The capacity factor of concentrated solar is 21.8 percent, which we will round to 20 percent. Therefore, on average, we need 120,000 MW_th concentrated solar installed to supply 24,000 MW_th thermal energy or 1,000 MW_t power over twenty-four hours. Each particular day's intermittent energy production is different. On extremely sunny days, we are producing much, much more energy than we can use, and we have a low return on assets. On extremely cloudy days, we are not producing enough energy; hopefully, we have enough storage to make it until tomorrow. Over the long-run average, we can last until tomorrow.

What about the short run? Given the energy we captured today, will we last until tomorrow? Weather comes in runs, with several extremely sunny days in a row followed by several cloudy days in a row. We can now see that we will need a storage system in addition to our concentrated solar farm capacity. When we have a five-day run of extremely sunny days from Monday through Friday, we need to save the daily excess of 96,000 MW_th for Saturday, or for a day perhaps two weeks from now or more. Otherwise we are throwing away energy that went into our calculation using the average capacity factor. Therefore, the storage system has to be quite big, holding at least 500,000 MW_th, and it has to be of long duration, lasting fourteen days or more.[5] In the end, an engineer and a financier will work out the probabilities of weather, the length of weather runs, generation capacity, and storage capacity. They then balance those costs against the loss of revenue from stopping or slowing production and the increased costs of operating the chemical plant below optimal efficiency.

The broad point is that using concentrated solar process heat for industry requires multiple solar farms and large storage. We can compare this against the relatively much simpler solution of using dispatchable nuclear process heat.

Using nuclear, we need to build one 1,000 MW_t nuclear plant,[6] which has an average rounded 90 percent capacity factor. We must determine how to keep the chemical plant running for the remaining 10 percent of the time. Nuclear is dispatchable. Therefore, we can time the nuclear plant's maintenance and refueling with the chemical plant's maintenance and reconfiguration, allowing us to place an effective 100 percent capacity factor in the table.[7]

	Concentrated solar	Gen IV
Installed daily energy capacity	120,000 MW_th	24,000 MW_th
Average daily energy capacity	24,000 MW_th	24,000 MW_th
Nameplate power capacity	5,000 $_{MWt}$	1,000 MW_t
Capacity factor	20%	100%
Average power capacity	1,000 MW_t	1,000 MW_t
Nighttime power capacity	0 MW_t	1,000 MW_t
Amount storage needed	500,000 MW_th[8]	0 MW_th
Extra assets	storage	–

TABLE 43. Process heat comparison summary.

With a high-intermittency, low-capacity energy source, we have to build more structures, resulting in a low return on assets. This is easily seen as we compare the installed versus average energy between the two technologies. Concentrated solar will have process heat applications, but because it is only semi-dispatchable, those applications will be in specific niches rather than in broad application.

HEAT PUMPS

A heat pump is a lot like a refrigerator. To cool a building using a refrigerator, place the evaporator inside (open the refrigerator door) and the condenser outside (move it such that the coils on the back are outside the building), and plug the gaps between the inside and the outside. Cooling a building works even on a hot summer's day, requiring more pumping the hotter it is outside. To heat a building, rotate the refrigerator until the components are reversed: the condenser inside and the evaporator outside. *Voilà!* This ability to heat the building works even in the winter, requiring more pumping the colder it is outside. A refrigerator is simply a heat pump with an insulated box.

For a family house or a commercial building, the best choices for heat pump installation are dependent on climate, building size, and the type of thermal reservoir (also known as the outside of the building). The thermal reservoir type may be ambient air, soil, stone, ponds, rivers, or groundwater. I advise all homeowners to investigate your building's suitability with a reputable heat pump installation service.

The temperature below ground does not change much over the course of a year. Several meters of soil are an effective insulation against the temporary cold of winter or heat of summer. Below-ground heating can be improved by putting down several tons of rocks or concrete slabs; such large heat sinks can store summer heat for winter or vice versa. Therefore, many designs place the thermal reservoir under parks and lawns and even under buildings.

Modern heat pumps often use supercritical carbon dioxide as the working fluid, giving new meaning to the concept of carbon capture and storage!

NUCLEAR POWER CONCEPTS

NUCLEAR ENERGY IS released when an unstable atomic nucleus fissions into two smaller, unequal fission products and two or three neutrons. The two fission products are unequal; one is a light atom, such as ^{95}Kr, and the other is a heavier atom, such as ^{137}Ba.[1] The neutrons are very fast, but they can be slowed down by the use of a moderator, such as helium, carbon, regular water, or heavy water.[2]

Although many atoms can undergo fission, there are only three practical fuels. Uranium-233 (^{233}U) is created from thorium and is usable in either a thermal or a fast reactor. Uranium-235 (^{235}U) is created by enriching uranium ore from 0.7 percent ^{235}U to 3 to 5 percent ^{235}U, and it is usable in a thermal reactor. Plutonium-239 (^{239}Pu) is created from the remaining 99.3 percent of uranium ore (^{238}U) and is usable in a fast reactor.

When a neutron collides and sticks to an atom of these nuclear fuels, the nucleus becomes unstable and has a high probability of fission, each reaction producing a few neutrons. A reactor is built so that the neutrons are reflected inward toward the fuel to sustain a chain reaction. When many fission reactions take place simultaneously, an appreciable level of heat can be extracted for power generation.[3]

A GEN II/III NUCLEAR PLANT

Most Gen II/III nuclear plants use 3 to 5 percent enriched ^{235}U solid fuel pellets bundled into fuel assemblies. They use water as a neutron moderator, producing thermal spectrum neutrons, and they incorporate safety features learned since the accidents at Three Mile Island. They operate with water as the primary coolant under very high pressures of 70–160 atmospheres.[4] Many designs have passive safety features to prevent meltdowns or other accidents. Gen II/III nuclear plants commonly are built using one of three reactor designs: a pressurized water reactor, a boiling water reactor, or a heavy water reactor.

The drawback of the Gen II/III plant is the price tag. Price is driven by the plant size and on-site construction, which, in turn, are driven by the systems necessary with a water coolant design requiring pressurized containment vessels.

The most modern Gen III plant is probably the NuScale Power Small Modular Reactor. This reactor deals with the pricing issue by grouping together from one to twelve small reactors together within a single containment shell. The power generation overhead is shared between these small reactors. Even if all twelve reactors fail, the plant can cool all twelve to a safe point without external power because each individual reactor is so small.

A GEN IV NUCLEAR PLANT

Gen IV nuclear plant designs are differentiated by their choice of coolants and choice of a thermal or fast neutron spectrum. They can use any of the usual nuclear fuel options (^{233}U, ^{235}U or ^{239}Pu) either as a liquid or solid. Many designs operate with salt or metal as a coolant at atmospheric pressure—bypassing the need for pressurized containment. In a meltdown accident, the fuel is captured by the salt or metal coolant. Examples of these plant types are the Terrestrial Energy Integrated Molten Salt Reactor and the Moltex Energy Stable Salt Reactor.

Salt and lead eutectics are inert; they do not burn, degrade, disassociate, or explode. Salt and lead do boil, but the boiling points of the salts and lead eutectics under consideration are at extremely high temperatures, well above achievable peak reactor temperatures. The reactors are designed with drain tanks or dilution tanks. These tanks rely on unpowered automatic temperature-sensitive valves to permit drainage or dilution to keep the coolants far from the boiling point. The lack of water prevents the possibility of Chernobyl-like steam explosions or Fukushima Daiichi-like hydrogen explosions.

Two other design types use sodium or helium as a coolant. Helium is inert but requires high pressures in the coolant loop. Sodium is a low-pressure coolant but will burn in the presence of air or water. The interested reader will readily find further background at the Generation IV International Forum (gen-4.org), the International Atomic Energy Association (iaea.org), the US Energy Information Agency (eia.gov), or Wikipedia.

Gen IV is designed to be constructed with easy-to-install modules. Gen IV salt or metal coolant reactor designs need containment vessels but do not require *pressurized* containment vessels. Lack of pressurized containment allows off-site manufacture and on-site assembly, making them smaller and cheaper to construct than Gen II/III reactors.

BUILDING TITANIC-SCALE GEN III NUCLEAR POWER BY 2050

Creating enough reactors to fully decarbonize electricity would simply require a concerted effort of building 144 plants each year until coal and natural gas plants are finished.[5] France has demonstrated that Gen II/III reactors take an average of 7.22 years to build, but the most common build time is 5.5 years. In addition, France started nine reactors each in 1975 and 1979 and four or more reactors in seven other years. The lesson from France is that, with determination and ability, a country can build a lot of Gen II/III reactors in a short period of time.

China has more recently demonstrated the same. China's Gen II/III reactors take an average of 6.03 years to build, but with strong project management, we may reasonably expect construction to take less than six years. In addition, China started nine reactors in 2009 and eight in 2010, illustrating an ability to build at scale. China confirms the French lesson: With determination and ability, a country can build a lot of Gen II/III reactors in a short period of time.

During the country's nuclear power plant construction period in the 1970s, the French GDP per capita was around $8,500, which is roughly equivalent to $36,000 in 2020. There are approximately twenty-four nations today with a larger GDP per capita than France had in the 1970s. The Chinese started most of their nuclear power plants around 2010, when their GDP per capita was around $4,000, roughly equivalent to $4,800 in 2020. There are well over eighty nations with economies at the same GDP per capita.

	France	China
Period	Late 1970s	Late 2000s
Maximum starts (reactors/year)	9	9
Mean rate(reactors/year)	7.22	6.03
Mode rate(reactors/year)	5.50	5.58
GDP per capita during build-out	$8,500	$4,000
2020 dollars	$36,000	$4,800
Number of nations with similar GDP	24	80

TABLE 44. Economies able to support nuclear power plants.

If the twenty-four nations with a GDP equivalent to 1970s France each start a mere six nuclear power plants at the rounded French and Chinese rates of six years each, then the world can start 144 plants a year. Assuming 1,000 MW power plants with a 90 percent capacity factor, we can reach a fully low-carbon replacement of electricity generation by 2046.

Year	Plants started	Cumulative plants finished	Percentage toward converting global electricity to low-carbon power
2020	144	–	–
2021	144	–	–
2022	144	–	–
2023	144	–	–
2024	144	–	–
2025	144	–	–
2026	144	144	4.8%
2027	144	288	9.6%
2028	144	432	14.4%
2029	144	576	19.2%
2030	144	720	24.0%
2035	144	1,440	47.9%
2040	144	2,160	71.9%
2042	144	2,304	81.5%
2045	144	2,880	95.9%
2046	144	3,024	100.7%

TABLE 45. Ability of world to scale nuclear construction.

Of course, the much richer countries are capable of building more than six reactors a year, and nearly eighty countries are capable of building at least one or more reactors a year. When we add in the secondary contributions from hydropower, geothermal, wind, solar PV, and concentrated solar, the picture becomes even better. By 2042, 80 percent nuclear and 20 percent renewables could replace fossil fuel electricity generation. Although the demand for electricity will grow, we can clearly see that Gen II/III power supported by renewables is a viable path toward a low-carbon future.[6]

BUILDING TITANIC-SCALE GEN IV
NUCLEAR POWER EVEN MORE EASILY

If Gen II/III is a viable path, then Gen IV nuclear plants are an even better path. The interesting Gen IV reactors have major advantages compared to Gen II/III that enable a quicker, cheaper build-out. The main construction advantage over Gen II/III plants is the factory-based manufacture due to the smaller size needed for operation at atmospheric pressure. This makes Gen IV both simpler to construct and safer to operate.

Atmospheric-pressure operation enables Gen IV containment vessels and piping to be reduced from 20 cm (8 inches) to a few millimeters-thick steel. Seals, pumps, gaskets, and plugs are more effective at lower internal pressures. Because of this lack of thickness, the reactor vessel size also decreases—in turn, simplifying casting, machining, and welding.

The much smaller size of Gen IV reactors permits manufacture in factories. Factory-based production raises the quality of manufacture, increases the rate of manufacture, and promotes learning. The decreased size permits shipment of whole reactors on a single flatbed truck. Plant construction speeds up because the most important component, the reactor, is installed rather than built.

The inert salt and lead eutectic coolants of Gen IV are much safer than water coolant, and so safety systems become cheaper and simpler.

The higher temperatures of Gen IV permit a higher Carnot efficiency in electricity generation and higher energy efficiency, with direct use of the coolant in process heat applications. These higher efficiencies expand the market and lower operational costs while raising a plant owner's benefits.

NUCLEAR WASTE ONLY NEEDS TO BE STORED FOR 300 YEARS

You may believe that nuclear waste needs to be entombed for several millennia. However, those time frames are only needed if society continues to pursue current disposal methods. Current Gen II/III commercial nuclear power plants work with a wide range of uranium fuels, from 1 to 5 percent, with different waste profiles. Although any given reactor's percentages will be different from what is described here due to differences in fuel composition and reactor design, the qualitative result will be very similar. Looking at 3 percent ^{235}U, the problem is caused by the remaining 97 percent of impurities, mostly ^{238}U. The uranium fuel lifecycle starts with this 3 percent ^{235}U fuel going into the reactor. What comes out of the reactor are 2 to 3 percent fission products plus whatever ^{235}U did not undergo fission. The 97 percent remainder is ^{238}U contaminated with long-lived transuranics.

The fission products are highly radioactive isotopes of xenon, krypton, cesium, strontium, iodine, barium, and many others, of which the maximum half-life is approximately thirty years. These fission products will decay to background levels within 300 years or so.

The long-lived transuranics are somewhat radioactive isotopes of plutonium, neptunium, americium, californium, and others. It is difficult to separate uranium from transuranics; however, it is easy to separate the fission products from the uranium and transuranics. After separation, place the uranium and transuranics back into the reactor where they belong. The reactor will eventually convert these leftovers into fission products, generating electricity all the while.

We can solidify the separated fission products by melting them in silica to create a glass impervious to water. Then we bury the glass ingot for 300 years. Examples abound of humans storing even the most delicate of treasures, such as jewelry, paintings, ceramics, and other art for long periods of time. Burying a bulk lump of stone-like glass for 300 years is comparatively much simpler.

Input fuel	Output fuel	Decay time	Solution
3% ^{235}U	2%–3% fission products—Xe, Kr, Cs, Ba, I, etc.	decays to background in 300 years	bury for 300 years
97% ^{238}U	96% ^{238}U and remaining ^{235}U	same as went in	return to reactor
	1% transuranics—Pu, Am, Cf, etc.	irrelevant inside the reactor	return to reactor

TABLE 46. Disposal strategy for spent nuclear fuel.

Gen IV reactor designs simplify this separation process compared with their Gen II/III counterparts. In addition, the thorium fuel cycle can be substituted for the uranium fuel cycle in many Gen IV design types. One of the advantages of the thorium fuel cycle is that it inherently produces fewer long-lived actinides than the uranium fuel cycle, in turn reducing the complexity of the fission products separation step.

DO NUCLEAR PLANTS CROWD OUT WIND, SOLAR PV, AND CONCENTRATED SOLAR FARMS?

Carbon dioxide emissions is the problem causing climate change. A plant that cannot crowd out a high-carbon coal or high-carbon natural gas plant is not a plant worth building. We need power plants that go forth and kill coal plants.

Nuclear plants kill coal plants. If nuclear technology crowds out wind, solar PV, and concentrated solar technologies, then those technologies are not killing coal. If wind, solar PV, and concentrated solar farms are killing coal plants, then those technologies will not be crowded out by nuclear technology.

NUCLEAR WEAPONS PROLIFERATION

This is a political question, not one of technology or money, and it can only be solved politically. Civilian nuclear power generation has very little to do with creating nuclear weapons. In fact, one does not need to build a reactor to build a nuclear weapon.

^{235}U bombs need an enrichment facility, not a reactor, to separate the isotope ^{235}U from raw uranium ore. A civilian reactor competes for ^{235}U; building civilian reactors makes ^{235}U bombs more expensive.

^{239}Pu bombs need a plutonium factory, a reactor operated in short fuel cycles of a few weeks at low power. A civilian power reactor is designed for long, multiyear fuel cycles operating at high power. A civilian power reactor creates a mix of plutonium isotopes that hopelessly contaminate the plutonium from the perspective of creating a nuclear weapon. Operating a civilian power reactor at low power and low capacity factors (due to the frequent shutdowns to replace the fuel) would stick out like a sore thumb. Remember: The average reactor is generating at peak design power with 90 percent capacity factors and is expected to do better year-on-year, as would the more frequent replacement of components not designed for those conditions.

However, let us assume you are the leader of a state with nuclear weapons, such as the United States, France, United Kingdom, Russia, China, or India. Building another civilian nuclear power plant will not affect your existing nuclear weapons program. The political question before you is whether to dismantle your nuclear weapons, not whether to proliferate weapons based on your civilian power program.

Next, let us assume you are the leader of a state with nuclear power but without nuclear weapons, such as Belgium, Finland, Korea, Japan, Mexico, Slovakia, Sweden, or Ukraine. Neither your existing civilian nuclear power plants nor your ability to build another reactor will affect your (nonexistent) nuclear weapons program. The political question before you is whether to design and build nuclear weapons, not whether to build another power plant. These countries have already opted to not develop nuclear weapons.

Alternately, let us assume you are the leader of a responsible state with neither civilian nuclear power nor nuclear weapons, such as Chile, Kenya, Poland, or Vietnam. States without nuclear power need the assistance of those with it if they want to develop such plants. Therefore, there is leverage to guard against nuclear weapons development for states wanting to construct civilian nuclear power stations.

Last, let us assume you are the brutal dictator of a totalitarian state with neither a civilian nuclear power plant nor a nuclear weapon. You are determined to develop a nuclear weapon. Do you take the difficult route and spend a huge amount of money to build a civilian power reactor? This would draw the attention of the International Atomic Energy Agency and other interested parties. You would have to ship in 97 percent impure fuel that is closely tracked by the shipper, and you'd have to create a large power plant that can be monitored by satellite. A power plant would create plutonium that is too contaminated to use in a bomb. Or do you take one of

the easier routes and instead spend a much smaller amount of money, build
no reactor at all, and construct an enrichment (^{235}U bombs) facility or a
plutonium (^{239}Pu bombs) factory?

There are two states that have built nuclear weapons but have not built
a civilian nuclear power reactor. There are thirty states that have built civilian nuclear power reactors but have not built nuclear weapons. Seven out
of nine states with nuclear weapons built them before their first civilian
power reactor.

28 states built only civilian power reactors[*]	9 states with nuclear weapons[§]
Argentina	**5 built nuclear weapons before civilian reactors**
Armenia	
Bangladesh	China
Belgium	France
Brazil	Russia
Bulgaria	United Kingdom
Canada	United States
Czech Republic	
Finland	
Germany	**2 built civilian reactors before nuclear weapons**
Hungary	
Iran	India
Italy	Pakistan
Japan	
Kazakhstan	
Lithuania	**2 built nuclear weapons, and have not built civilian reactors**
Mexico	
Netherlands	Israel
Romania	North Korea
Slovakia	
Slovenia	
South Africa	
South Korea	
Span	
Sweden	
Switzerland	
Taiwan	
Ukraine	

TABLE 47. Civilian nuclear power does not promote nuclear weapons proliferation.

Sadly, it is probable that states with nuclear weapons will keep them with or without the existence of civilian nuclear power plants. The ability of states with civilian nuclear power to build nuclear weapons is a greater reflection of their wealth and engineering capacities than a reflection of how their electricity is generated.

Politics, not the civilian electricity mix, is behind the motivation of states to build nuclear weapons. Given that seven of the nine states possessing nuclear weapons built them before they built their civilian nuclear power, there appears to be little crossover between civilian and military nuclear technologies.[9]

TERRORIST ATTACKS ON NUCLEAR POWER PLANTS

Let us assume you are instead a nihilistic terrorist.[10] If you can raise the several billion dollars to build a civilian nuclear power reactor, you have probably already started down the path of a much cheaper plutonium factory or a uranium enrichment facility. However, the infrastructure you need is still at the nation-state level, and so your local brutal dictator has probably already usurped your physical plant.

If, instead, you attack a civilian nuclear power plant, fuel reprocessing plant, or fuel in transport, you likely did so for one of three possible reasons. If you did it to gain uranium, you are quite stupid; there isn't enough enriched uranium in a processing plant to make a bomb. If you did it to gain plutonium, you are quite stupid; the ^{239}Pu you need is useless for a bomb because it is contaminated with ^{240}Pu, ^{241}Pu, and ^{242}Pu. If you did it to make a dirty bomb, you are somewhat stupid; the public may panic because this sort of bomb disperses radioactive material, but it will not be more destructive than a standard explosives attack.[11]

However, perhaps the nuclear plant itself is your target. Aside from the public's fear, it is difficult to blow up the plant, and it is not sited near large groups of people. An oil refinery or natural gas pumping station is a much more exposed and satisfying target: It does go boom,[12] it is not as well protected, and it is much closer to large groups of people.

The Fukushima Daiichi accident shows how tough nuclear plants are. Fukushima Daiichi suffered a magnitude-nine earthquake—destructive energies far beyond the capabilities of a sub-nation-state terrorist group. Then a 14 m (45 feet) tsunami struck—again, destructive energies far beyond the capabilities of a sub-nation-state terrorist group. Even at that point, the plant was in good condition. In a submerged power plant surrounded by the complete devastation of a natural disaster, it still took four and a half hours for the first core to be uncovered, twenty-four hours for the first hydrogen explosion to occur, and forty-eight hours for the second core to be uncovered. Only the total devastation of the infrastructure around the plant prevented its staff from saving the reactors. A nation-state would have responded with police action well within these time periods.

Terrorism, nuclear or otherwise, is a policing issue, not a technical or money issue of nuclear power.

NUCLEAR REACTORS ARE NOT NUCLEAR BOMBS

Nuclear power plants are incapable of a nuclear explosion producing a mushroom cloud and flattening a city. A blowup explodes, but a meltdown oozes. At worst, a Gen II/III nuclear reactor can reach very high temperatures, beyond the melting point of the zirconium cladding of the fuel, at which point the low-density fuel will melt into ooze. As the ooze melts the reactor vessel, more ooze is created, further lowering the fuel density. As the fuel becomes less dense, the chain reaction stops, the temperature lowers, and the ooze freezes in place. This low-neutron-activity, low-density ooze is the exact opposite of what is needed for a nuclear bomb.

A Gen IV reactor during normal operation is cooled by molten salts or molten metals, essentially pre-oozed and always ready for a potential accident. In an accident, the low-density fuel oozes directly into this pre-oozed coolant, trapping the now-lower-density fuel and the fission products for later cleanup. Often these reactors have a temperature-triggered drain for the ooze to exit the reactor core, shutting the reactor down automatically. Drainage occurs naturally under gravity, and so external power is not required to shut down the reactor. As the fuel exits the reactor, the chain reaction stops, the temperature lowers, and the ooze freezes in place. This ooze, again, is the exact opposite of what is needed for a nuclear bomb.

A nuclear bomb works by using plastic explosives to compress high-density fuel into a very small volume of a few cubic centimeters[13] in a fraction of a second. This raises the density such that all the fuel undergoes a chain reaction in microseconds. The high-neutron-activity, high-density compressed fuel needed for a nuclear explosion is the exact opposite of what is occurring under either normal or accident conditions within a nuclear power plant.

A reactor can never explode like a nuclear bomb. However, it is possible for a nuclear bomb to meltdown like a reactor. If the bomb's fuel were to reach critical mass too slowly to explode, it would melt together, creating a very hot ooze, which would melt the bomb, eventually cooling into a pile of slag.

In short, a reactor would need both highly pure fuel and an explosive shockwave to raise the density such that the critical mass was within a volume to cause a nuclear explosion. It has neither the 90 percent pure fuel nor the thousands of tons of plastic explosive lying about that would be needed to compress a volume[14] of tens of cubic meters (hundreds of cubic feet) containing the fuel and tons of steel and concrete comprising the reactor walls and piping, to create a critical mass of fuel.

	Gen II/III reactor	Gen IV salt/lead reactor	Bomb
Nuclear fuel	97% impure, low-density fuel	97% impure, low-density fuel	90% pure, high-density fuel
Fuel surroundings	many tons of steel and concrete	many tons of steel and concrete and a few tons of salt/lead	high temperature shockwave produced by a few kilograms of plastic explosive
Accident result	low-density ooze of fuel, zirconium, steel, and concrete	low-density ooze of fuel, salt/lead, steel, and concrete	high-density plasma
Density direction	expanding to ever-lower density stopping chain reaction	expanding to ever-lower density stopping chain reaction	compression to high density starting 100% chain reaction
Timescale	hours or days to solidify	minutes or hours to solidify	microseconds to explode
Aftermath	possible steam/hydrogen explosion	not explosive	fireball, shockwave, mushroom cloud

TABLE 48. Nuclear plants have neither the pure fuel nor the high density to create nuclear explosions.

PERSPECTIVES ON SAFETY

We must have a sense of proportion. The tragedies of the Chernobyl and Fukushima Daiichi accidents are very painful for their victims. However, our society makes trade-offs every day: Smoking tobacco and burning coal are legal activities, despite both causing death from lung cancer, heart failure, and asthma.

There is also the matter of fear. More people are afraid of flying than driving, but airplanes are much safer than automobiles. Likewise, nuclear power is the safest form of electricity generation there is. People drown in hydropower dam failures. Maintenance workers fall from wind turbines 150–200 m (500–650 feet) tall. Construction workers fall off roofs while installing residential solar PV. However, more people would die without electricity at all; hospitals would cease to function; vaccines and other drugs require refrigeration; food requires cooking, pasteurization, and refrigeration; people can die from heat stroke without air conditioning or freeze to death without heat. Nuclear is the safest[15] electricity-generation technology—safer than coal, natural gas, hydropower, geothermal, wind, solar PV, and concentrated solar.

THE THREE MILE ISLAND ACCIDENT

The 1979 Three Mile Island accident illustrates a nuclear plant's strong safety systems and strong designs:

- Zero people were hospitalized.

- Zero people died from radiation exposure.

- Zero people died from radiation-induced cancer.

- The baseline lifetime increased cancer risk for Harrisburg residents is 0 percent.

Deaths[16]	Type	Name
1,800 to 25,000	dam failure	Machchhu dam failure
50	oil tanker explosion	Betelgeuse incident
49	natural gas explosion	Warsaw gas explosion

TABLE 49. Comparison with 1979 energy accidents.

THE CHERNOBYL ACCIDENT

Chernobyl No. 4[17] is not representative of commercial nuclear power. This reactor was not designed as a civilian power reactor but as a military plutonium factory. As such, it was designed and built without a containment building. The staff were poorly trained and poorly prepared. The plant was not designed or built to then-current Soviet standards of civilian power operation. It was literally "a plant which fell well short of the safety standards in effect when it was designed and even incorporated unsafe features."[18]

Even so, the human casualties[19] of sixty-two dead and a few thousand thyroid cancers were relatively light for such a major industrial accident caused by the steam explosion and fire:

- One hundred thirty-four people were hospitalized due to acute radiation syndrome (ARS).
  - Twenty-eight people died from ARS within three months.
  - Nineteen people died from ARS or other causes (including accidents) over the next seventeen years.
  - Eighty-seven people recovered from ARS.

- Fifteen died from thyroid cancer.
- In total, 4,000 to 9,000[20] people have contracted or are subject to a higher-than-expected chance to suffer thyroid cancer over the remainder of their lives.[21]

There have been 6,000 diagnosed cases of thyroid cancer related to the accident. These cases are inclusive of the normal rate of thyroid cancer incidence and of subclinical diagnoses that would not be noticed under normal scrutiny.

In response to the Fukushima Daiichi accident, Germany has shut down half of its nuclear plants while keeping open or building new high-carbon coal plants. Germany continues to shut down low-carbon nuclear plants. German coal plants have killed 4,350 people each year due to cancer and respiratory diseases. By continuing to use coal power, Germany has essentially inflicted a Chernobyl-size accident on Europeans every year—and promises to continue this destruction until 2038.

A 30-km radius exclusion zone is monitored for excessive radionuclides around the Chernobyl site. This area is now a wildlife reserve, thriving due to the absence of large numbers of humans.

Deaths	Type	Name
177	mine fire	Kinross mine accident

TABLE 50. Comparison with 1986 energy accidents.

THE FUKUSHIMA DAIICHI ACCIDENT

Fukushima Daiichi Unit 1, Unit 2, and Unit 3, and the fuel pool of Unit 4 are representative of commercial nuclear power in the West. These Gen II reactors were designed to produce civilian power, had proper containment buildings, and were operated by a well-trained staff. The human casualties were relatively light for a major industrial accident caused by hydrogen explosions:

- Thirty-nine people were injured.
- One person who died from lung cancer five years after the accident was awarded worker's compensation, although the court explicitly stated it was unlikely to be due to the accident.
- An estimated 1,600–2,200 people died from fear and panic.[22]

The baseline lifetime cancer risk for Fukushima residents has increased approximately 1 percent. For comparison, lifetime lung cancer risk is approximately 1 percent for nonsmokers and approximately 22 percent and 12 percent, respectively, for male and female smokers.

Deaths	Type	Name
120	pipeline oil fire	Nairobi Pipeline fire
26	mine explosion	Sukhodilska-Skhidna July 2011 coal mine accident
11	mine elevator collapse	Bazhanov July 2011 coal mine accident

TABLE 51. Comparison with 2011 energy accidents.

HYSTERIA AND MORTALITY ESTIMATION

The 1952 Great Smog of London gives us perspective on the killing capabilities of burning coal. It also shows us what proper scientific research looks like compared with hysteria and fear mongering. In 1952, from December 5th to December 9th, a freak weather pattern trapped coal-generated smog over London. Lasting a mere four days, the result was an immediate death toll of some 4,000 people, and some 100,000 people were injured. Due to lung diseases such as bronchopneumonia, the death toll remained higher than average for several months after the incident. Modern estimates place the death toll at around 12,000 from the event.[23] This research has been conducted with access to death certificates and gravesites.

The governments of the Soviet Union and, since its dissolution, Ukraine, Belarus, and Russia have conducted similar scientific studies on the impact of the Chernobyl accident. The United Nations World Health Organization and the International Atomic Energy Agency (IAEA) have also conducted such studies. As stated previously, the sad consensus is that approximately sixty-two people have died as a known result of the disaster,[24] and in excess of the background rate, an additional 4,000–11,000 will contract cancer over their lifetime.

There are other estimates of mortality rates, many of which include deaths in countries other than Ukraine, Belarus, or Russia. Are the following estimates believable?

- The Union of Concerned Scientists estimates 50,000 excess cancers, 25,000 of which will be terminal.

- Greenpeace estimates over 200,000 excess deaths between 1990 and 2004 and 270,000 excess cancers, resulting in 93,000 terminal cases.

- Alexey Yablokov, Vassily Nesterenko, and Alexey Nesterenko estimate 985,000 excess deaths between 1986 and 2004.

- Helen Caldicott, a noted antinuclear campaigner, has estimated that millions died as a result of the accident and has alleged that a dark conspiracy exists to cover up these deaths by governments, the United Nations, the IAEA, and the nuclear power industry.

I find these other estimates to be unbelievable. The simple fact is that, unlike the 1952 Great Smog of London, the corroborating evidence does not appear to exist. There should be piles of documentation from death certificates and hospital admissions. There should be a vast number of extra gravesites and hordes of cancer patients. Finally, the number of grieving relatives does not appear to match these estimates. We can rule out a dark conspiracy of thousands of officials across tens of countries to destroy documents and hide graves. Over three decades have passed[25] since the Chernobyl accident; we can stop the hysteria and rely on science and rational assessment of the actual documentation to establish the historical outcome.

The Fukushima Daiichi accident is approximately a decade old now. Official and peer-reviewed reports of an excess cancer rate vary between very low and undetectable. We can cease frightening people. We can rely on science and experience from the Chernobyl accident to provide proper context to the residents of Fukushima Prefecture.

FUSION OPTIMISM

There is a group of advocates for fusion power who are very optimistic, even to the point that they suggest society delay fission-based nuclear and focus on fusion-based nuclear power. The optimism about fusion should be tempered with the knowledge that fusion does depend on breakthrough science, novel engineering, and undeveloped technologies.

Fusion technology is still decades away because scientific progress is still required. The technologies do not all exist, and placing fusion success within a ten-year time frame may or may not be possible.

Fusion is difficult. All fusion approaches involve input energy to start the reaction. Many approaches need to heat hydrogen to 100,000,000°C in an extremely confined space and keep the hydrogen in this state for long periods of time. It is so hard that a "long period of time" is considered to be a few milliseconds. The bulk of the design is simply starting and sustaining the fusion reaction. The remainder of the design is similar to fission: dealing with containment and removing heat energy via a coolant to generate power. Fusion has many open scientific inquiries and painful engineering problems, over decades-long timelines and costing multiple billions of dollars to resolve. The bottom line is that fusion is difficult to start and harder to sustain.

In contrast, fission is easy. Unlike fusion, no energy input is needed. The reaction simply starts itself when control rods are physically moved. The simplest approach has been to mine uranium ore, purify it, and without enriching it, place it in a moderated environment. As with fusion, the rest of the design deals with containment and removing heat energy via a coolant to generate power. Two such reactors are the Chicago Pile 1, which was constructed in 1942 and used natural uranium interleaved with graphite moderator, and the CANDU (Canadian Deuterium Uranium), which was constructed in 1968 and uses natural uranium surrounded by heavy water moderator. Simply take enough uranium and mix with the proper proportion of heavy water or graphite in a shielding vessel that can transfer heat, and—*voilà*—you have a fission reactor. Fission is remarkably simple to start and effortless to sustain.

Humanity cannot make a bet as to when fusion will graduate from the laboratory experimental level. After graduation, we still need to develop prototype reactors and demonstration reactors. Then there is the time to complete a fleet of commercial fusion reactors. Although humanity will probably develop fusion power within the next 100 years, it simply is not a technology as simple as fission power. Fusion will remain costly experimental science and is unlikely to become commercial electricity generation.

MARK JACOBSON'S *WHY NOT NUCLEAR?*

Mark Z. Jacobson has presented a policy argument against the use of nuclear power. His argument, taken from his presentation *Why Not Nuclear?*, is examined here.[26] The main slide from that presentation consists of the following five points:

NUCLEAR PRODUCES SIX TO TWENTY-FOUR TIMES MORE CO_2 PER KWH THAN WIND (LIFECYCLE COST)

Dr. Jacobson's first point appears incorrect: A quick check shows that the accepted values by the IPCC are 11 gCO_2/kWh for wind energy and 12 gCO_2/kWh for nuclear power. His determination of six to twenty-four times is due to calculating an opportunity cost of CO_2 produced while building nuclear power plants. Therefore, if nuclear truly takes ten to nineteen years to build, this determination is an interesting point.

NUCLEAR TAKES TEN TO NINETEEN YEARS TO BUILD VERSUS TWO TO FIVE YEARS FOR WIND OR SOLAR

The French and Chinese have shown that it is possible to build Gen II/III nuclear in 5.5–7.2 years, and it is clear that Gen IV nuclear is buildable in two to four years because they have simple design licenses,[27] are flexible in siting, are quick in planning, and only require on-site construction of an earthquake-grade foundation and a turbine hall. The reactor itself is manufactured in a factory, delivered on a truck, and installed rather than constructed on-site. Due to these timelines, I feel that both Dr. Jacobson's first and second points may be disregarded.

NUCLEAR COSTS THREE TO FOUR TIMES MORE THAN WIND OR SOLAR PV AND NUCLEAR TAKES TWO TO TEN TIMES LONGER TO OBTAIN ONE-THIRD TO ONE-FOURTH THE CO_2 SAVINGS PER DOLLAR OF WIND OR SOLAR

As we have learned, wind and solar PV plants need to be built in a minimum of triple the current rate—and even more, depending on our storage capacity, grid intelligence, and tolerable risk levels. Nuclear plants may cost three to four times what wind and solar PV farms cost, but we require three to four times the wind and solar PV farms as we do nuclear plants, so the costs even out.

Furthermore, a nuclear plant is an asset whose value can be depreciated over forty to eighty years, whereas the average wind or solar PV farm's value is depreciated over twenty-five to thirty years. Essentially, the nuclear plant will last two to three times as long. We may get value out of a Maggie Fry masterpiece, but Leonardo da Vinci's *Mona Lisa* has a greater staying power, a higher value.

Finally, although the costs are nearly equivalent, cost is irrelevant in comparing these plants. The real question is "Which technology has higher value?" Nuclear power is dispatchable power, has a high capacity factor, requires only a single-site generation plant, and lasts two to three times as long as a wind or solar farm does while costing about the same in total. Storage and grid upgrades are welcome but not necessary. Wind or solar energy, on the other hand, is intermittent energy, has a low capacity factor, requires multiple-site generation farms, and lasts half as long as a nuclear plant while costing about the same in total. These options cost more when we include the necessary storage and grid upgrades.

This fourth point on the time taken to realize emissions reductions per dollar depends on the previous two being correct. I believe that this calculation is incorrect, given the appropriate construction timelines and costs of nuclear power plants.

THE FIFTH POINT ON THE RISKS OF NUCLEAR

Dr. Jacobson cited the Intergovernmental Panel on Climate Change in claiming "robust evidence, high agreement" that nuclear leads to risks of weapons proliferation, meltdown, waste, and mining.

Weapons proliferation is a political issue, not a technical one. Countries with nuclear weapons already have them. Building a civilian nuclear power plant will encourage a country to make neither more nor fewer bombs. Countries with civilian nuclear power plants have pretty much already decided whether to build bombs or not; again, building another civilian nuclear power plant will not encourage a country to build bombs. Countries wishing to build nuclear weapons will try to either buy such a weapon, steal such a weapon, or set up a program to build such a weapon; they certainly will not go down the expensive path of civilian nuclear power with its nonexistent crossover technologies to build such a weapon.

The meltdown risk for Gen IV reactors is basically nonexistent. As Gen II/III reactors are phased out in favor of Gen IV, their risk will decrease. Training and vigilant regulators will keep Gen II/III meltdown risk low.

Waste risk is addressable by building Gen IV reactors, then using them to convert our many tons of long-lived high-level waste to a few kilograms of short-lived fission products.

Mining risk is quite low. A few small mines produce a few thousand tons of raw uranium ore per year. Uranium mining has a comparatively low geographic footprint and environmental impact. Using the thorium or plutonium fuel cycles reduce mining impact even further. In contrast, hundreds of large mines produce many millions of tons of coal. Natural gas is produced in the billions of cubic meters across tens of thousands of gas wells. Although an individual wind turbine, solar PV panel, or concentrated solar reflector requires less steel, concrete, and other materials than a nuclear power plant, the aggregate materials needed for an equivalent-capacity renewables farm is greater than for a single nuclear plant.

NET-ZERO METERING

ELECTRICITY HAS HISTORICALLY been amortized into a marginal rate over the following costs:

- · The cost of generation (construction, maintenance, and operations)
- · The cost of transmission (construction and maintenance)
- · The cost of distribution (connection, installation, and maintenance)

Today, residential or commercial buildings are generating their own electricity, usually with solar PV, and their owners want to sell it back to the grid. This microgeneration is intermittent, with peak supply at noon when demand is average, and therefore problematic for the grid. Electricity is consumed as it is produced, and that overproduction cannot be consumed. Alternatively, we can state that electricity needs to be produced as it is consumed and no earlier.

Unfortunately, many solar advocates and politicians believe the grid is similar to a bank, where you may put energy in first and then remove it later. Banking is the incorrect analogy. The grid, unlike a bank, does not store energy, does not accumulate energy interest, and cannot lend energy before it is generated. Because electricity is produced and consumed in the moment, storage is needed, and someone must pay to construct it.

The correct analogy to energy storage is physical storage space, such as garages or self-storage units. Someone must pay to construct, maintain, and operate physical storage. Therefore, fees are charged for use of the building. Likewise, energy storage must be constructed, maintained, and operated.

Let us examine these questions:

CONGESTION

If the householder is relying on the grid-level storage, should they be charged extra for depositing at peak supply and withdrawing at peak demand?

SELF-RELIANCE

If the householder is not relying on the grid for storage, why do they need to be connected to the grid?

CONVENIENCE

If the householder is only relying on the grid for power at critical moments, shouldn't they be willing to pay a premium for critical-moment power?

DISCRIMINATION

If the householder is selling power to the grid, should they sell at the same wholesale price as other generators? Should they sell at the same retail price as they bought at?

GRID OVERHEAD COSTS

If the householder is a generation source, should they pay their share of capital and maintenance for transmission and installation? Should they be required to keep a spinning reserve (extra juice when a lamp turns on)? Should they pay safety costs for the grid, such as tree trimming, buried cable location, and safety courses?

TIMING

Is the contract uneven? Does the householder appear to be forcing a sale at the time (peak supply) of their convenience rather than at the time of the grid's peak demand?

EXCHANGE SERVICE

The utility's grid enables the household to sell, not to the utility, but to other households. Should the utility collect a broker's fee for that transaction?

	Net-zero metering	Self-reliant	Fair exchange
Storage analogy	bank	attic	storage unit
Storage supplier	utility	householder's batteries	utility
Finance	someone else pays	householder owns storage	householder rents storage
Householder buys at	retail	retail	retail
Householder sells at	retail	–	wholesale
Contract externalities	unfair to other ratepayers	fair	fair

TABLE 52. Comparison of possible microgeneration market structures.

In the net-zero metering scenario, the householder buys and sells electricity at the same price. The time value of the generation is not priced for this consideration, and grid costs are not assessed. The household installs a solar PV array, collects fees during the middle of the day when electricity is over-generated, underused, and creating problems for grid stability. It is a double-subsidized ride for the household; everyone else pays for this externality.

In the self-reliant scenario, the householder purchases a solar PV array and a battery. Any extra juice is sent to the battery; to the rest of the grid, the household looks very energy efficient. No externalities are imposed on others.

In the fair exchange scenario, the householder purchases a solar PV array but needs the grid to provide juice at night or during high demand. The householder pays less in utility fees and sells to the grid on the same turns as other generators, paying for maintenance and management, but imposes no externalities on others.

NET-ZERO METERING IS A SLUSH FUND

Some readers may believe that utilities are evil corporations just trying to milk the ratepayer and taxpayers for a buck. Therefore, by forcing the utilities to subsidize microgeneration via net-zero metering, we are getting some of our own back. If this is your belief, then you should be vociferously against net-zero metering. The wrong way to bring a utility to heel is to use opaque slush funds such as net metering. If you believe in subsidies for microgeneration, use government to sponsor that directly. Keep utilities under simple, easy-to-observe rules and far, far away from the large-scale potential for abuse of complex subsidies.

THE ELECTRIC GRID IS NOT LIKE THE INTERNET

ELECTRICITY IS NOT LIKE a series of data packets, and the electric grid is not like the internet. Electricity is immediate, and generation must meet the demand load instantaneously. The grid has very little storage. At each stage, the energy degrades: losses due to wires and transformers, losses due to charging and discharging storage, etc.

Although the internet is fast, it is not immediate. Data on the internet does not have to meet demand instantaneously; it is allowed to slow down and send the data when bandwidth frees up. The internet is dominated by storage: Bytes on disk, bytes in memory, and even bytes in transit are stored as a packet when received on each router node and then forwarded as it moves from computer to computer. The information does not degrade; if a bad packet is detected, it is simply resent.

If we ship Arizona data to New York, it will get there eventually. It is sent packet by packet, resent when errors occur, and delayed by seconds, minutes, or even hours when congestion occurs. When the data arrives, it is stored in memory or on a disk, ready for consumption at any point in time.

If we ship Arizona sun energy to New York as electricity, the demand in New York has to be the same as the Arizona generation. The electricity is sent in one shot (i.e., if a blip occurs, the energy is lost). The grid has to be preconstructed to handle the immediate load. A delay means a brownout or blackout in New York. Rerouting to mitigate a too-small grid requires *other* preconstructed grids.

Maybe we can ship Arizona sun to St. Louis and St Louis wind to New York. In this case, the St. Louis leg still has to meet the timing and demand load of New York. As the Arizona leg is backing St. Louis, it also has to meet the timing and demand load of New York, but the extra stopover has muddled the pure New York demand signal, decreasing grid responsiveness and increasing grid instability.

Maybe we can pre-transmit and store the electricity—but where do we store it? New York uses, on average, 17,000 MWh every hour. If New York had a Bath County Pumped Storage Station, eighty-four minutes of the state's energy can be stored. However, power is not used on average. At the peak, New York is consuming 41,079 MW, and Bath County can only generate 3,003 MW, a major gap.[1]

Finally, whether there is a single big wire connecting Arizona and New York or a series of big wires at each stage, along each length of wire, energy is being degraded and lost due to transmission.

A comparison between analog electricity and digital information will only remain a bad analogy. It will never become an engineering pathway.

ACKNOWLEDGMENTS

I WISH TO THANK the following people for their friendship, encouragement, and assistance in writing this book: David J. Lindstrom, Maureen McCabe, Dr. Thomas M. Dunn, Ryan J. Boyle, George S. Rieg, Catherine Corcoran, and Joe Longo.

I have deep appreciation for the staff at Greenleaf Book Group: Lindsey Clark, Claudia Voldman, Nathan True, Brian Phillips, Jen Glynn, Justin Branch, Corrin Foster, and Tiffany Barrientos.

NOTES

PREFACE

1. During preparation for publication, the IEA published new data for 2018. It is not any more meaningful to us to examine 24,738,920,000 MWh per year. That is why the concepts introduced by this book will remain timeless—rounding a few numbers up or down does not change the fundamental insights presented. International Energy Association (IEA) estimates of global electricity demand, https://www.iea.org/world.

2. A year has close to 365.25 days, which accounts for the extra leap day every four years.

3. Total energy also includes transportation, industrial, commercial, and residential categories. The percentages change across reports depending upon the definitions of what is or is not included in a category.

4. The United States Energy Information Agency (EIA) estimates the average American home consumed 10.9 MWh in 2018.

5. IEA estimates Belgium consumed 88,600,000 MWh in 2017, https://www.iea.org /countries/Belgium.

6. IEA estimates, 2017; https://www.iea.org/data-and-statistics.

7. Cement production is one of the most energy-intensive industrial activities. Industrial ammonia is, similar to cement, highly energy-intensive to produce and is the primary ingredient of fertilizer. Global famine would occur without fertilizer from Haber process ammonia. EIA, https://www.eia.gov/todayinenergy/detail .php?id=11911.

CHAPTER 1

1. Note that the multiplier 84 is over the twenty-year time frame.

2. Other fossil fuels, such as oil, do not contribute much to electricity generation; oil is used mostly for transportation. Fossil fuel electricity generation is primarily from coal and natural gas.

3. Air Quality Life Index, https://aqli.epic.uchicago.edu/.

4. J.P. McBride, R. E. Moore, J. Witherspoon and R. Blanco, "Radiological impact of airborne effluents of coal-fired and nuclear power plants," *Science* 202, Issue 4372 (December 1978): 1045-1050, https://doi.org/10.1126/science.202.4372.1045.

5. Actinides are the elements along the bottom of the periodic table: actinium, thorium, protactinium, uranium, neptunium, plutonium, americium, *etc*. A subset of the actinides are the transuranics, elements past uranium starting with neptunium. All of the actinides have radioactive isotopes.

6. IEA (2016), Energy and Air Pollution, IEA, Paris, https://www.iea.org/reports/energy-and-air-pollution.

7. American Lung Association, https://www.lung.org/clean-air/outdoors/what-makes-air-unhealthy/toxic-air-pollutants.

8. US emissions estimates from 2015 are extrapolated to world emissions. EIA, 2018, https://www.eia.gov/energyexplained/natural-gas/natural-gas-and-the-environment.php.

CHAPTER 2

1. One million people is approximately 0.013 percent of the global population; alternatively, 1 percent of the global population is approximately 76 million people.

2. Energy Star is a US program that provides energy efficiency ratings for products and services, https://www.energystar.gov/.

3. *Externality costs* are indirect costs imposed upon others. The classic example is the pollution your neighbor emits by burning leaves. The neighbor is dumping carbon dioxide into the air for free, but you pay the cost in a less healthy environment and more medical bills. https://en.wikipedia.org/wiki/Externality.

4. Webcast debate: "The World Needs a Nuclear Renaissance," Stanford ENERGY, https://youtu.be/uuJodGvyLzM.

5. Hunter S. Thompson, "The Pain of Losing," ESPN.com, https://www.espn.com /espn/page2/story?page=thompson/041109.

6. UNEP (2014), Assessing Global Land Use: Balancing Consumption with Sustainable Supply. A Report of the Working Group on Land and Soils of the International Resource Panel. Bringezu S., Schütz H., Pengue W., O'Brien M., Garcia F., Sims R., Howarth R., Kauppi L., Swilling M., and Herrick J., https://www.resourcepanel.org/reports/assessing-global-land-use.

7. Bernard L. Cohen, http://www.phyast.pitt.edu/~blc/book/.

8. IEA data, 2017, https://www.iea.org/data-and-statistics; UN HDI data, 2017, http://hdr.undp.org/en/content/table-1-human-development-index-and-its -components-1.

9. *The Magic Washing Machine*, https://youtu.be/BZoKfap4g4w.

CHAPTER 3

1. Prof. David N. Ruzic, the IllinoisEnergyProf, offers a great series of tutorials on energy technologies, https://www.youtube.com/channel/UCKH_iLhhkTyt8Dk4dmeCQ9w.

2. It is unfortunate that direct heat use is rarely separated out in the official accounting.

3. IEA data, 2017, https://www.iea.org/data-and-statistics.

4. *Ibid;* Note that the biofuels and waste resource is not represented in Figure 14 as it was in Figure 13. This is because those emissions are often (incorrectly in the author's view) attributed to the industry that originated the feedstock such as forestry, agriculture, or municipalities. Although inclusion would have a minor effect on the percentages, the basic proportions will hold true.

5. IEA data, 2017, https://www.iea.org/data-and-statistics.

6. Improvements to the efficient use of and generation of heat are a different matter, as will be explained later in this chapter.

7. The other and biofuels category contains the contribution of incinerating waste, and also geothermal and concentrated solar due to their extremely small percentage-wise contribution. Oil, more useful in transport, is used very little in electricity generation. Hydro and nuclear appear to have swapped proportions from the total energy chart. This is because the efficiency of nuclear is reduced by the conversion from heat to electricity. Using heat directly avoids efficiency losses from conversion; IEA data, 2017, https://www.iea.org/data-and-statistics.

8. *Ibid.*

9. To create a 1,000 MWe (subscript e for electric) power plant, at a 40 percent Carnot efficiency, we need to supply 2,500 MWt (subscript t for thermal) power. Thus a 1,000 MWt power plant is much smaller than a 1,000 MWe power plant. Each power plant has its own unique Carnot efficiency rating, as they each have their own input and output temperatures. A 40 to 50 percent Carnot efficiency is quite a good rating. Unless explicitly stated, MW power refers to electric power.

10. More properly, energy is the quantitative property that is transferred to an object in order to perform work on it or to heat it. Power is the rate at which the quantitative property is transferred. As an example, water at the top of a hill has potential energy, which can flow down the hill and release kinetic energy. Power is the rate at which the water's flow is converted to kinetic energy.

11. An average thunderbolt contains several billion joules of energy, which over the milliseconds of discharge is trillions of watts of power. Outside of rare applications like DeLorean time machines, a large energy resource is not necessarily a practical power resource. https://en.wikipedia.org/wiki/Lightning.

12. To convert watts to joules per second, or more commonly from joules to watt-hours, we need to remember there are 3,600 seconds in an hour, 24 hours in a day, and 8,766 hours in a year.

13. The mode is the most common value. In this case, continuous power has a mode of 1,000 MW, because 8,766 of the 8,766 hourly measures in the year are at 1,000 MW. The fast draining battery has a mode of 0 MW, because 8,765 of the 8,766 hourly measures are at 0 MW. The 1-hour drain-and-fill battery has two modes—discharging at plus 1,000 MW and charging at minus 1,000 MW.

14. Comparison of the energy capacity of different electricity generation types should be over the course of a year as that incorporates the full seasonal and daily variation in both generation and consumption. This comparison is constrained by the requirement to keep the balance between power generated and power consumed during every second of the year.

15. Such a power plant can be designed with a higher nameplate capacity as calculated by the capacity factor and including allowances for intermittency, as will be discussed later on in this book.

CHAPTER 4

1. Baseload power and peak power are often conflated with dispatchable power. Baseload and peak are after-the-fact measures of electricity generation in a given period of time. Plant owners want high-baseload generation near the peak demand, as they can then run their power plants flat out, optimizing their return on assets. However, a grid operator and society only care about matching the current demand load with the current generation capacity. Thus, the concept of interest is the ability to dispatch either more or less electricity.

2. The Texas ERCOT grid has a high penetration of wind generation. In 2009, the maximum rate change recorded a change in generation of minus 619 MW over ten minutes, equivalent to minus 3,714 MW per hour. However, this event only lasted thirty minutes before its direction was reversed and generation recovered. In terms of total installed wind capacity, this was a 7 percent ramp over the ten-minute period. That much variability strains both grid operations and equipment.

3. Many weather conditions, such as cold and warm fronts, are highly correlated weather across extremely large areas, the size of 100,000 km^2 (38,600 square miles). For example, a third of the continental United States, the Midwest, can experience a single blizzard for days at a time. A large hurricane or typhoon may cover up to eight degrees of latitude, an area some 890 km (550 miles) in diameter.

4. *Critical* in this sense has no connotation of importance. Critical simply indicates that the appliance will fail without power. You may view the ability to watch television as unimportant, but the electric company does not know if they are powering a residential television displaying a game show or powering vital equipment in a hospital's intensive care unit. A television used to monitor an intensive care patient is vital equipment.

5. By definition, high-intermittency plants cannot be dispatched. Intermittent plants cannot be designed to run on demand at high outputs for long periods of time. Intermittency is a function of wind, clouds, or other natural phenomena, not of the plant's design.

6. The raw data shows how little the capacity factor changes year to year. EIA capacity factor data, https://www.eia.gov/electricity/monthly/epm_table_grapher .php?t=epmt_6_07_b.

7. Interestingly, athletics does impact electricity demand. Grid operators pay close attention to major sports events such as the World Cup, Olympics, World Series, etc., when billions of televisions and radios create an enormous simultaneous demand load.

8. Analogously, even though the crowd at a sports stadium on game day is quite large, the average crowd is quite low, perhaps even zero, because most days are not game day. The average is effectively useless in predicting crowd size at any given moment. The game day average crowd size is more interesting, but the mode is the better predictor of crowd size—the first mode is the usual attendance on game day, and the second mode is zero on game-less days. The average, or mean, is a mathematical construct and does not imply the usual or steady-state condition. In a population of five one-year-old children and one seventy-nine-year-old grandfather, the average age is fourteen years, and the mode is one. In this case, the average does not help in predicting the age of an individual. The mode, the most common value, is often a better predictor of individual members of a population than the average.

9. Texas wind, ERCOT data, January 2019, http://www.ercot.com/gridinfo/generation.

10. The capacity factor is used as a prorated average for intermittent energy resources in several of the scenarios outlined. All the scenarios are very much in the ideal world; the intermittent energy scenarios are much worse in reality. To further complicate calculations, intermittent resources are often partially on, rather than fully on or fully off. As dispatchable power can be run constantly over time, dispatchable capacity factors lead to results representative of the real world.

11. Do not confuse an efficiency increase with reduced intermittency. Efficiency improvements can capture more wind or sun energy, but only if the wind is blowing or the Sun is shining. Intermittency is a function of wind, clouds, or other natural phenomena, not the efficiency of the plant design.

12. North American Electric Reliability Corporation, "Accommodating High Levels of Variable Generation," April, 2009, https://www.nerc.com/pa/RAPA/ra/Reliability%20Assessments%20DL/Special%20Report%20-%20Accommodating%20High%20Levels%20of%20Variable%20Generation.pdf.

13. Wind and solar seasonal maximums and minimums are specific to different geographies. The seasonal maximums and minimums cited here are prevalent across the United States. Any particular wind or solar farm site will have seasonal variation in addition to hourly or daily variation. EIA, https://www.eia.gov/todayinenergy/detail.php?id=20112.

14. Overcapacity equals a low return on assets due to the opportunity costs incurred. In other words, the money spent on an idle asset could be used to construct something more productive.

15. Coal boilers can take four to nine hours to reach the heat necessary for electricity generation. The average natural gas turbine takes between ten to sixty minutes to reach the heat threshold necessary for electricity generation. To keep time-to-generation low, coal and natural gas plants use fuel to remain warm. Thus, even when on standby, they still produce greenhouse gas emissions. Peter Kokopeli, Jeremy Schreifels, Reynaldo Forte (2013). "Assessment of startup period at coal-fired electric generating units." U.S. Environmental Protection Agency, Office of Air and Radiation, https://www.epa.gov /sites/production/files/2015-11/documents/matsstartstsd.pdf.

16. Electricity grids vary widely in area and do not always follow political borders. An excellent resource to visualize these can be found at https://www.electricitymap.org/.

17. Robert Hargraves, in his book *THORIUM: Energy Cheaper Than Coal*, estimates that a paired wind/natural gas plant generates 44 percent more greenhouse gas than an always-on natural gas plant.

18. A hurricane with a diameter of 890 km (550 miles) covers an area 90 percent the size of Texas.

19. Return on investment and return on assets are basically interchangeable concepts in the context of this book. Return on assets is more relevant to our comparison as project financing is a sunk cost, and the returns are to be made through operating the asset.

20. Winter and summer are reversed seasons on the southern hemisphere of the planet. However, the great majority of the world's population lives in the northern hemisphere. In any case, reversing the argument still leaves energy production at a minimum when it is needed most.

21. An individual solar farm's capacity factor is influenced by latitude and prevailing weather. Equatorial solar farms will have higher capacity factors than polar solar farms. Solar farms in sunny deserts will have lower intermittency than those sited on foggy seacoasts.

22. A problem that can be solved with money is not a problem. Who pays is not a problem of money, but rather a problem of politics.

CHAPTER 5

1. IEA data shows that electricity growth was 4 percent in 2018, up from 2.5 percent growth in 2017, and is projected to grow approximately 2.1 percent annually over the next twenty years. A projection for electricity growth depends upon assumptions of economic growth, substitution of current fossil fuel use with electrification, substitution of current industrial energy use with process heat and the penetration rate of electricity to support currently under-served populations. That analysis is beyond the scope of this treatise. However, the astute reader will immediately realize that electricity generation will grow a great deal due to both reducing high-carbon energy resources through electrification and supporting a growing and under-served world populace; IEA (2019), World Energy Outlook 2019, IEA, Paris, https://www.iea.org/reports/world-energy-outlook-2019.

2. IEA (2019), Will pumped storage hydropower expand more quickly than stationary battery storage?, IEA, Paris, https://www.iea.org/articles/will-pumped-storage -hydropower-expand-more-quickly-than-stationary-battery-storage.

3. The decimal places are correct, that is 0.038 percent (not 0.38 percent, nor 3.8 percent). Global storage capacity is not just small, but tiny: 1.4 percent of chemical battery storage translates to 0.00053 percent of total energy. The other problem is that the useful size of storage depends a great deal on the storage timescales needed. The amount of storage needed over the time frame of seconds does not need to be huge. The amount of storage needed over days, weeks, and seasons does need to be Atlas-size.

4. In reality, the Bath Country Pumped Storage Station is not operated on such a twenty-four-hour cycle. Bulk storage is not the station's primary purpose; it provides other services to the grid, such as frequency regulation, voltage support, and other ancillary services.

5. Note that the *ideal* Bath County of this scenario stores an extra 12,000 MWh over the *actual* Bath County facility.

6. Average energy (versus average power during operation) needs to take into account the capacity factor. Scenarios presented in this book usually make the unrealistic simplification that power can be averaged out evenly using the capacity factor.

7. Of course, at the real-world efficiency of 79 percent, it consumes energy, thus netting negative. Storage only shifts the time of dispatch; and the frictions of storage are always net negative.

8. The major function of the grid is to provide electric power. Ancillary services keep the grid stable. https://en.wikipedia.org/wiki/Ancillary_services_(electric_power).

9. Belgium generated 88,630,000 MWh of electricity in 2018, IEA, https://www.iea .org/countries/Belgium.

10. Global installed storage, growth, and technologies; IEA, https://www.iea.org/topics /tracking-clean-energy-progress.

11. G.P. Hammond, S.S. Ondo Akwe, S. Williams, "Techno-economic appraisal of fossil-fuelled power generation systems with carbon dioxide capture and storage," *Energy*, 36, Issue 2 (2011): 975-984, https://doi.org/10.1016/j.energy.2010.12.012.

12. An example of a long cycle-time fuel is wood. The Joseph C. McNeil Generation Station claims biomass sustainability with a twenty-five to thirty-year carbon-neutral cycle to support a population of 42,000 people. Wood may be a renewable fuel, but it is difficult to perceive of such a long cycle as sustainable. The Good Stuff: *Can a City Run on 100% Renewable Energy?*, https://youtu.be/zKhzVcHrWH4.

13. Repeal of the Clean Power Plan; Emission Guidelines for Greenhouse Gas Emissions from Existing Electric Utility Generating Units; Revisions to Emission Guidelines Implementing Regulations, https://www.federalregister.gov/documents/2019 /07/08/2019-13507/repeal-of-the-clean-power-plan-emission-guidelines-for- greenhouse-gas-emissions-from-existing.

14. Ammonia can be used as a fuel. In 1822, David Gordon patented an ammonia engine. In 1869, Emile Lamm sold streetcars powered by ammonia commercially. In 1943, wartime Belgium used ammonia hybrid buses. In 2007, the University of Michigan developed an ammonia-gasoline internal combustion engine hybrid. http://www.nh3car.com/.

CHAPTER 6

1. Boundary Hydroelectric Project, http://www.seattle.gov/light/Boundary/.

2. Run-of-river is essentially an underwater wind turbine. No dam is needed, merely the flowing water of a river.

3. Dr. David Mackay's *Sustainable Energy—Without the Hot Air;* https://withouthotair.com/.

4. The maximum tides in the Baltic Sea are 17–19 cm (6.7–7.4 inches). Medvedev, I.P., Rabinovich, A.B. & Kulikov, E.A., "Tidal oscillations in the Baltic Sea," *Oceanology* 53, (2013): 526–538, https://doi.org/10.1134/S0001437013050123.

5. The Banqiao and Shimantan dams in Zhumadian Province, China, failed in 1975. It is estimated that 26,000 died immediately due to failure of the dams, with 145,000 estimated dead in the aftermath of the accident. https://en.wikipedia.org/wiki/Banqiao_Dam.

CHAPTER 7

1. Hellisheidi Geothermal Power Station, https://www.on.is/en/about-us/power-plants/.

2. IEA (2020), Geothermal, IEA, Paris, https://www.iea.org/reports/geothermal.

CHAPTER 8

1. Centrale Nucléaire de Cattenom (Cattenom Nuclear Power Plant), https://www.edf.fr/groupe-edf/producteur-industriel/carte-des-implantations/centrale-nucleaire-de-cattenom/presentation.

2. France generated 479,230,000 MWh of electricity in 2018. IEA, https://www.iea.org/countries/France.

3. NuScale, each plant taking 36 hectares (90 acres), https://www.nuscalepower.com/technology/technology-overview.

4. IEA data, 2017, https://www.iea.org/fuels-and-technologies/nuclear.

5. International Atomic Energy Agency (IAEA), *Nuclear Power Reactors in the World*, Reference Data Series No. 2 (Vienna, Austria: IAEA, 2019), https://www.iaea.org/publications/13552/nuclear-power-reactors-in-the-world.

6. The distinction between Gen II and Gen III plants is a matter of industry perspective but is unimportant to an examination of energy policy. Most Gen II/III plants use high-pressure water as a coolant—the high pressure allowing higher temperatures for greater Carnot efficiency—in contrast to Gen IV plants that can achieve high temperatures at low pressure.

7. The Carnot theorem calculates the efficiency of any general heat engine. To improve the efficiency of the turbines, the coolant needs a large temperature change. One way to do this is to lower the temperature of the coolant before reintroduction to the reactor—this is why many Gen II/III power plants have the iconic cooling towers. The other way to improve efficiency is to raise the temperature of the coolant exiting the reactor, creating higher-grade heat, as in Gen IV.

8. Overnight cost is the cost of construction including materials and labor. It excludes financing costs such as interest payments on loans. The overnight cost is useful in comparing the costs of different plants constructed at different time periods. As an example, plant A may be expensive, but built in a time of low interest rates and so have a low total cost; plant B may be inexpensive, but built in a time of high interest rates and so have a high total cost. By ignoring the financing component, we can determine that plant B is the cheaper design to use in a future project, irrespective of what the future financing cost will be.

9. Simplified chart based on MIT's *The Future of Nuclear Energy in a Carbon-Constrained World*, https://energy.mit.edu/research/future-nuclear-energy-carbon-constrained-world/.

10. A eutectic mixture has a lower melting point than the melting points of its constituents.

11. Molten salt is simply salt heated past its melting point. In contrast, sodium chloride (NaCl, table salt) is frozen at temperatures below 800°C. Contemplate the fact that you are sprinkling frozen salt next time you add it to your food. Common eutectics of salt are mixtures of sodium fluoride, beryllium fluoride, lithium fluoride, potassium chloride, and sodium chloride.

12. Gen IV reactors still require containment structures. Gen II/III reactors require 20 cm (8 inch) thick pressure vessels and difficult welds. A Gen IV containment vessel only has to support the mass of itself and its contents at the operating temperature range and so only needs to be several millimeters (a fraction of an inch) thick with easy welds. To see a common design and cost analogy, compare a stove-top espresso maker made of thick stainless steel that contains steam (100°C) at a few atmospheres pressure to a French press coffee maker made of a thin glass jar that contains boiling water (100°C) at one atmosphere.

13. "The End of Coal in Ontario," https://www.ontario.ca/page/end-coal#section-3.

14. The volume of high-level waste is about 22,000 m³ (a cube 28 m ≈ 92 feet on each side). Hannah Paterson (2019), https://nda.blog.gov.uk/2019/08/02/how-much-radioactive-waste-is-there-in-the-world/.

15. Moorburg emits approximately 8,500,000 tons carbon dioxide per year in exchange for 1,730 MW power, https://de.wikipedia.org/wiki/Kohlekraftwerk_Moorburg.

16. Pastoria emits approximately 1,600,000 tons carbon dioxide per year in exchange for 780 MW power. EIA, https://www.eia.gov/beta/electricity/data/browser/#/plant/55656.

17. The count of 623 includes both operational and retired reactors.

18. In a nuclear reactor, there is a *slightly enriched (0–5 percent)* slow-fissioning fuel and *nothing to raise the density of* the ooze in an accident. A nuclear bomb works by using plastic explosives to compress a *highly enriched (90 percent-plus)* fast-fissioning fuel in a fraction of a second, *raising the density* such that all the fuel undergoes a chain reaction in microseconds, with no time to create an ooze. It is easy to see that a nuclear reactor and a nuclear bomb are completely disparate technologies. See the appendices for a more detailed comparison.

19. German coal kills 1,860 internal to Germany and 2,500 external each year. Of course, these deaths are estimated statistically; the same as has been used to estimate the cancer incidence due to the Chernobyl accident. The brutal point is that after shutting down half their nuclear reactors, in essence Germany creates a Chernobyl-sized disaster every year by continuing to burn coal; Dave Jones, Julia Huscher, Lauri Myllyvirta, Rosa Gierens, Joanna Flisowska, Kathrin Gutmann, Darek Urbaniak, Sarah Azau, *Europe's Dark Cloud: How Coal-Burning Countries Are Making Their Neighbours Sick*, http://env-health.org/IMG/pdf/dark_cloud-full_report_final.pdf.

20. Pushker A. Kharecha, Makiko Sato, "Implications of Energy and CO2 Emission Changes in Japan and Germany after the Fukushima Accident," *Energy Policy* 132 (2019): 647-653, https://doi.org/10.1016/j.enpol.2019.05.057.

CHAPTER 9

1. Wind farms have hundreds of separate turbines; one of the most common manufactured is the GE 1.5 MW turbine. A farm with 1,000 MW of nameplate capacity requires 667 of these turbines.

2. Presented are typical values for wind turbines; each design has different trade-offs and resulting values.

3. Roscoe Wind Farm, https://en.wikipedia.org/wiki/Roscoe_Wind_Farm.

4. Gansu Wind Farm Project, https://en.wikipedia.org/wiki/Gansu_Wind_Farm.

5. Common problems on megaprojects, https://www.nytimes.com/2017/01/15/world/asia/china-gansu-wind-farm.html.

CHAPTER 10

1. Which weighs more, a ton of lead or a ton of feathers? Analogously, the spacing of panels does not mitigate shadow. A small dark shadow and a large lighter shadow still prevent the same amount of sunlight from reaching vegetation, leading to the same lack of growth in total.

2. Topaz Solar Farm, https://en.wikipedia.org/wiki/Topaz_Solar_Farm.

3. North American Electric Reliability Corporation, "Accommodating High Levels of Variable Generation," April, 2009, https://www.nerc.com/pa/RAPA/ra/Reliability%20Assessments%20DL/Special%20Report%20-%20Accommodating%20High%20Levels%20of%20Variable%20Generation.pdf.

CHAPTER 11

1. In concentrated solar, a common coolant is a salt eutectic of sodium nitrate ($NaNO_3$) and potassium nitrate (KNO_3), which matches the 250–600°C temperatures reached by this technology.

2. Solana Generating Station, https://en.wikipedia.org/wiki/Solana_Generating_Station.

CHAPTER 12

1. Remember, peak generation occurs at noon—supply outstrips demand. Peak demand occurs after the sun sets—demand outstrips supply.

2. During the winter day, concentrated solar provides heat; during the winter night, the heat pump uses either the stored hot coolant or the ambient environment. During the summer, the heat pump cools, using the concentrated solar collector reconfigured as the condenser's radiator.

CHAPTER 13

1. These are median values from the IPCC 2014 report, Table A.III.2, Lifecycle emissions. Schlömer S., T. Bruckner, L. Fulton, E. Hertwich, A. McKinnon, D. Perczyk, J. Roy, R. Schaeffer, R. Sims, P. Smith, and R. Wiser, 2014: Annex III: Technology-specific cost and performance parameters. In: *Climate Change 2014: Mitigation of Climate Change. Contribution of Working Group III to the Fifth Assessment Report of the Intergovernmental Panel on Climate Change* [Edenhofer, O., R. Pichs-Madruga, Y. Sokona, E. Farahani, S. Kadner, K. Seyboth, A. Adler, I. Baum, S. Brunner, P. Eickemeier, B. Kriemann, J. Savolainen, S. Schlömer, C. von Stechow, T. Zwickel and J.C. Minx (eds.)]; Cambridge University Press, Cambridge, United Kingdom and New York, NY, USA, https://www.ipcc.ch /site/assets/uploads/2018/02/ipcc_wg3_ar5_annex-iii.pdf.

2. Direct process heat does not lose energy due to the inefficiencies of converting heat to electricity, transmitting electricity over potentially hundreds of kilometers, and then converting electricity back to heat. Gen IV designs have great installation flexibility, permitting new process heat in older industrial parks. Siting plants a few hundreds or thousands of meters away from factories allows hot coolant to be piped with low thermal losses. An industrial plant using concentrated solar would have to be moved to the concentrated solar site.

3. The valuation attributes of dispatchability, storage requirement, and smart-grid requirement appear to be duplicate attributes. However, these are different concepts, each driven by intermittency. Dispatchability allows grid operators to dial up or dial down a generation source in response to the demand load. Dispatchability is an inherent attribute of each generation technology. Storage is one mitigation strategy for intermittent energy farms and is a money-insoluble, technical problem. A smart grid is a second mitigation strategy for intermittent energy and a money-soluble, nontechnical problem.

4. Using electricity to produce carbon-neutral fuels permits longer-term storage. As we also have to solve the problem of low-carbon transport, those fuels are reserved for the transport sector.

5. Hirth, Lion, "The Market Value of Variable Renewables," *Energy Policy* 38, (2013): 218–236, https://doi:10.1016/j.eneco.2013.02.004.

6. Currently intermittent energy sources do not pay for the external costs of storage and grid improvements. This subsidy should be made explicit whether or not society continues this financial policy.

7. Remember the distinction between conservation and reduction. Energy conservation, the more efficient use of energy, is good for society and individuals. Energy reduction, the rationing of energy, is bad for both society and individuals.

8. *Renewable Electricity Futures Study (Entire Report)*, National Renewable Energy Laboratory. (2012). Renewable Electricity Futures Study. Hand, M.M.; Baldwin, S.; DeMeo, E.; Reilly, J.M.; Mai, T.; Arent, D.; Porro, G.; Meshek, M.; Sandor, D. eds. 4 vols. NREL/TP-6A20-52409. Golden, CO: National Renewable Energy Laboratory. https://www.nrel.gov/analysis/re-futures.html.

9. The acronym RE-ETI stands for Renewable Electricity—Evolutionary Technology Improvement. NREL's report has a 90 percent penetration scenario, but throughout the report the 80 percent scenario is used as the comparable, and so the 80 percent scenario is used here.

10. Relative nameplate capacity assets probably drop in high-proportion nuclear scenarios, but I do not have enough insight into the constraints of the NREL report to calculate relative asset size. It could be quite a huge impact, as nuclear's capacity factor of 90 to 95 percent is nearly double that of coal and natural gas's 55 percent, triple that of wind's 35 percent, and quadruple that of solar's 25 percent.

11. *Ibid.*

12. French electricity generation is 71 percent nuclear, and the impact is immense— France is a leader in reducing greenhouse gas emissions from electricity generation at 0.69 tons CO_2 per capita (compared to the European Union average of 2.19). IEA data, 2017, https://www.iea.org/data-and-statistics.

CHAPTER 14

1. Fossil fuel power plants need to pay a tax to use the biosphere as a waste dump. The tax reflects the indirect costs imposed on the rest of society, but it can also be set at a punitive rate to discourage behavior that causes these costs. How the taxes are spent is irrelevant. The taxes are meant to change behavior. Investment in low-carbon research and pilot plant subsidies should be allotted separately, independent of the revenue these taxes bring in.

2. Excess German wind has caused issues for Polish and Czech grid managers, which are not big enough to absorb the energy, https://www.thegwpf.com/germanys-renewables -revolution-threatens-neighbours-with-grid-collapse/.

3. IEA data on French, German, and EU renewables and emissions. Total emissions per capita, including transport and other sectors in addition to electricity generation, are as follows: Germany—8.24 tons/capita, EU—6.26 tons/capita, France—4.35 tons /capita. IEA data, 2018, https://www.iea.org/statistics/.

CHAPTER 15

1. Dr. David Mackay's TedEd talk, https://youtu.be/E0W1ZZYIV8o.

2. Emissions from agriculture, forestry, land use, and other sectors must also be addressed. However, they are beyond the scope of this book.

APPENDIX D

1. Heat capacity is also volume dependent—a given substance has the properties of heat capacity in terms of heat per mass, and density in terms of mass per volume. Large-scale thermal storage is both heavy and large.

2. Pumped hydro is approximately 97 percent; chemical batteries are approximately 1.4 percent; and compressed air is approximately 1.2 percent of storage. All the other types comprise the remaining balance of 0.4 percent. IEA (2019), "Will pumped storage hydropower expand more quickly than stationary battery storage?," IEA, Paris, https://www.iea.org/articles/will-pumped-storage-hydropower-expand-more -quickly-than-stationary-battery-storage.

3. Using an alkaline earth metal such as beryllium, magnesium, or calcium, we can store two charges per atom, reducing the volume. However, the energy density remains similar as the mass rises, because those metals are heavier than lithium.

4. Grid-level storage for power management such as frequency control and voltage regulation is a different story than grid-level bulk storage as described in the main text. Of course, coal and natural gas plants should be driven out of existence with a carbon tax. It is also likely that $0.02/kWh is incorrect in a given geography and date; however, at other prices the calculations result in similar conclusions. The essential insight remains valid even if using the slightly higher-priced low-carbon alternatives of dispatchable hydro, geothermal, or nuclear power plants.

5. The difference in the cost of solar PV compared to the cost of concentrated solar shows that exporting intermittency to the grid is highly beneficial to the owner of a solar PV farm. By extension, the same argument applies to wind, although in a less clear-cut manner due to political, financial, and geographical differences.

6. IRENA (2019), Innovation outlook: Smart charging for electric vehicles, International Renewable Energy Agency, Abu Dhabi, https://www.irena.org/publications/2019/May/Innovation-Outlook-Smart-Charging.

7. The real-world demand load of each of these cities is far higher than a mere 900 MW.

APPENDIX E

1. The United States' National Fire Protection Association recommends that hospitals have ninety-six hours (four days) of backup power, which applies to emergency systems only. The recommendation is not advocating that hospitals replace the grid power with their own power plants. Backup power is a temporary power source to deal with emergencies that affect power lines and other infrastructure, not a business-as-usual scenario that would replace the grid. https://www.nfpa.org/codes-and-standards/.

2. *A problem that may be solved with money is not a problem.* However, creating wind or sunlight when needed is not a problem that can be solved with money.

3. The Heliogen company can produce concentrated solar process heat as high as 1,500°C. https://heliogen.com/.

4. The subscript t indicates thermal power, rather than the usual electric power.

5. Note that the thermal storage system is not 100 percent efficient and suffers from both the minor energy losses of pumping coolant and the major energy losses of thermal radiation. The longer the coolant is in the storage tank, the more heat will be transferred to the tank. No storage tank can be perfectly insulated; the tank will be cooled by the ground on which it sits and the air that surrounds it.

6. Note that a thermal plant, using heat directly, is less than half the size of an electric plant, as the electric plant has to account for converting thermal power to electric power via the Carnot efficiency calculation.

7. Ten percent of the 365 days in a year is thirty-six days. Assuming a quarterly maintenance cycle allows us to apportion the maintenance in four nine-day periods synchronized with the chemical plant's quarterly maintenance cycle. Another solution is to construct two smaller 500 MW$_t$ reactors, allowing the chemical plant half-capacity operation during maintenance periods of one reactor. Of course, we can schedule a concentrated solar energy farm's maintenance at a convenient time also. However, we cannot schedule the weather—thus we may use up super-sunny days during maintenance, only to return to online operations during a *dunkelflaute*.

8. This is the amount of storage needed for fourteen days backup.

APPENDIX F

1. Further examples of the light atoms are isotopes of krypton, strontium, zirconium, or technetium; the heavier atoms are isotopes of iodine, xenon, cesium, or barium.

2. *Thermal* is the word nuclear scientists use to describe slow neutrons. Thermal neutrons have a high probability of causing an atom to fission because they spend more time closer to the nucleus when colliding, but they might not have enough energy to stick. Fast neutrons are very energetic and are, well, faster than the slow thermal neutrons. Fast neutrons have a high probability of causing an atom to fission because they collide more forcefully, but they might bounce off because they hit the nucleus so hard.

3. The number of fission reactions consumes several grams worth of uranium over a day. Reactor loads are on the order of several tons (depending on reactor size) of uranium every two to three years.

4. Water at 1 atmosphere boils at 100°C. The high pressures allow water to achieve temperatures around 300°C. The higher temperature permits a greater Carnot efficiency to be achieved.

5. Junji Cao, Armond Cohen, James Hansen, Richard Lester, Per Peterson, Hongjie Xu, "China-U.S. Cooperation to Advance Nuclear Power," *Science* 353, Issue 6299 (August 2016): 547-548, http://doi.org/10.1126/science.aaf7131.

6. How feasible is 144,000 MW a year? The coal industry, with plants similar in size and complexity to Gen II/III, brought 1,122,781 MW online over the last thirteen years, an average of 86,000 MW a year, with a peak of 94,000 MW in 2007. An expanding nuclear industry can easily construct 144,000 MW a year. This is a matter of will, not of feasibility. https://endcoal.org/global-coal-plant-tracker/summary-statistics/.

7. International Atomic Energy Agency, Country Nuclear Power Profiles, IAEA, Vienna (2018), https://www.iaea.org/publications/13448/country-nuclear-power-profiles.

8. Stockholm International Peace Research Institute, https://sipri.org/yearbook/2019/06/.

9. Nicolas Miller, "Why Nuclear Energy Programs Rarely Lead to Proliferation," *International Security*, Volume 42, Issue 2, (Fall 2017), https://doi.org/10.1162/ISEC_a_00293.

10. If you are indeed a terrorist, grow up and get a proper job so your family can stop being ashamed of you.

11. The explosion of a dirty bomb does the damage, not its radioactive material. In fact, radioactivity is easily measurable, thus radioactive hotspots can be quickly identified and cleaned up promptly.

12. The San Bruno pipeline accident created a wall of fire more than 300 m (1,000 feet) high, a crater 12 m (40 feet) deep, and an explosion that registered as a magnitude 1.1 earthquake. The explosion occurred in a residential neighborhood, killing eight and injuring fifty-eight. https://en.wikipedia.org/wiki/San_Bruno_pipeline_explosion.

13. A nuclear bomb core occupies a few cubic centimeters (a few hundred cubic inches).

14. The cores of a research reactor or a nuclear submarine are smaller, but also have correspondingly less fuel.

15. Hannah Ritchie, "What Are the Safest and Cleanest Sources of Energy?" https://ourworldindata.org/safest-sources-of-energy.

16. Comparison of Industrial Deaths, https://en.wikipedia.org/wiki/List_of_accidents_and_disasters_by_death_toll.

17. Only the No. 4 reactor suffered the accident. The remaining three reactors were kept online. No. 2 operated until 1991, No. 1 operated until 1996, and No. 3 operated until 2000.

18. International Nuclear Safety Advisory Group, The Chernobyl Accident: Updating of INSAG-1, INSAG-7, IAEA, Vienna (1992), https://www.iaea.org/publications/3786/the-chernobyl-accident-updating-of-insag-1.

19. United Nations Scientific Committee on the Effects of Atomic Radiation, http://www.unscear.org/unscear/en/chernobyl.html.

20. The International Agency for Research on Cancer has identified a slightly higher thyroid cancer rate of 11,000 in exposed populations and also an increased risk of eye cataracts in liquidation workers.

21. Thyroid cancer has a good prognosis for successful treatment. *Over the remainder of their lives* does not mean they will immediately suffer from such a cancer. For example, it could mean within one year for an eighty-year-old, or within eighty years for a one-year-old. https://en.wikipedia.org/wiki/Thyroid_cancer.

22. *New York Times*, 2015, https://www.nytimes.com/2015/09/22/science/when-radiation-isnt-the-real-risk.html; *Financial Times*, 2019, https://www.ft.com/content/000f864e-22ba-11e8-add1-0e8958b189ea; A 2017 study has shown, in stark terms, that the evacuations were not justified for the majority of the populations near Chernobyl No. 4 and Fukushima Daiichi.

23. Michelle L Bell, Devra L Davis, and Tony Fletcher, "A retrospective assessment of mortality from the London smog episode of 1952: the role of influenza and pollution," *Environmental Health Perspectives* 112 (2004):1, https://doi.org/10.1289/ehp.6539.

24. The number is approximate as the different source materials disagree by a few digits; sixty-two is the sum that has the broadest attribution to the accident. It encompasses those who died from acute radiation syndrome, cancer, or mental depression.

25. At this point in 2020, no ^{131}I (half-life of eight days) remains, and there is half the ^{90}Sr (half-life of twenty-nine years) and half the ^{137}Cs (half-life of thirty years) left. The most damaging radiation comes from these short half-life fission products; therefore, radiation readings must be about half of what they were in 1986.

26. *Why Not Nuclear?*, https://www.youtube.com/watch?v=yEHf5K9AQjY&t=938s.

27. A design license certifies the conceptual plant as designed. A site license certifies the specific plant to be constructed. Gen II/III plants have a near 1:1 ratio between these licenses, as the plants are built on-site. Most Gen IV plants will have a single design license that certifies the factory-built reactor and a site license for each of the site installations.

APPENDIX H

1. EIA data, 2019, https://www.eia.gov/electricity/state/newyork/index.php.

ABOUT THE AUTHOR

PRESTON URKA is passionate about environmental conservation and solutions to lower carbon-dioxide emissions and other pollutants. He was compelled to write this book to explain the titanic scale of our challenge. After writing this book, even he has had to change his perceptions of scale!

Preston received his BA in Mathematics and Chemistry from Southern Utah University in Cedar City, Utah. He received his MBA from Chicago Booth School of Business, University of Chicago in Chicago, Illinois.

Preston lives in Chicago's Ukrainian Village community, primarily because of the delicious cabbage rolls, bread, sausages, cheeses, and pastries to be found in the neighborhood. He is a devoted follower of the many Frontier League and Minor League baseball teams found near Chicago and the Windy City Rollers.